The Engineer's Contribution to Contemporary Architecture

ELADIO DIESTE

Series editors
Angus Macdonald
Remo Pedreschi

Department of Architecture
University of Edinburgh

The Engineer's Contribution to Contemporary Architecture

ELADIO DIESTE

Remo Pedreschi

Thomas Telford

Endorsed by

RIBA Publications

Published by Thomas Telford Publishing, Thomas Telford Ltd, 1 Heron Quay, London E14 4JD.
URL: http://www.thomastelford.com

Distributors for Thomas Telford books are
USA: ASCE Press, 1801 Alexander Bell Drive, Reston, VA 20191-4400, USA
Japan: Maruzen Co. Ltd, Book Department, 3-10 Nihonbashi 2-chome, Chuo-ku, Tokyo 103
Australia: DA Books and Journals, 648 Whitehorse Road, Mitcham 3132, Victoria

First published 2000

Also available from Thomas Telford Books

Engineer's Contribution to Contemporary Architecture - Antony Hunt. A. Macdonald ISBN 0 7277 2769 9
Engineer's Contribution to Contemporary Architecture - Heinz Isler. J. Chilton ISBN 0 7277 2878 4
Engineer's Contribution to Contemporary Architecture - Peter Rice. A. Brown ISBN 0 7277 2770 2

A catalogue record for this book is available from the British Library

ISBN: 0 7277 2772 9

© Remo Pedreschi and Thomas Telford Limited 2000

All rights, including translation, reserved. Except as permitted by the Copyright, Designs and Patents Act 1988, no part of this publication may be reproduced, stored in a retrieval system or transmitted in any form or by any means, electronic, mechanical, photocopying or otherwise, without the prior written permission of the Publishing Director, Thomas Telford Publishing, Thomas Telford Ltd, 1 Heron Quay, London E14 4JD.

This book is published on the understanding that the author is solely responsible for the statements made and opinions expressed in it and that its publication does not necessarily imply that such statements and/or opinions are or reflect the views or opinions of the publishers. While every effort has been made to ensure that the statements made and the opinions expressed in this publication provide a safe and accurate guide, no liability or responsibility can be accepted in this respect by the author or publishers.

Designed by Acrobat
Printed and bound in Great Britain by the Cromwell Press, Trowbridge, Wiltshire

Acknowledgements

Many people have helped in the preparation of this text. I would like to thank the following;

Eduardo Dieste, Gonzalo Larrembebere and Frederico Sanguinetti, from the office of Dieste y Montañez, Dr Jim Lawson, Fiona Mclachlan, Manuel Figeroa, Tracy Rammell, Gonzalo Larrembebere, Eduardo Dieste, Antonio Jiménez Torrecillas, Vincent del Amo, Carlos Clemente and Theresa Pedreschi. I have also enjoyed the collaboration and support of James Murphy of Thomas Telford and my fellow authors in the book series The Engineers Contribution to Contemporary Architecture, Dr Angus Macdonald, Dr John Chilton and Dr A Brown.

Dedication: This book is dedicated to the memory of Eladio Dieste 1917-2000.

Preface

This book is about the Uruguayan engineer, Eladio Dieste. From the first images that I saw of his work through to the completion of this text I have become increasingly enthralled by his ideas and his work and I am convinced that Dieste is one of the great engineers of the 20th century.

Dieste believed that technological innovation should be guided by a moral imperative, that the only justifiable position for development is the betterment of mankind, especially the more humble in society, whom Dieste felt deserved to be treated with great respect. This attitude can be traced throughout his original and innovative work. At present Dieste is best known for the primarily architectural qualities of his buildings, however, his designs demonstrate consummate technical skill and confidence and remove the boundary between architecture and engineering, between aesthetics and pragmatism. For Dieste these boundaries don't exist. At the heart of his work lies the expression of structural resistance through form, thus the architectural form and the structure become inseparable. Further more he believed that architecture, form and expression are also inseparable from economics, construction and structural efficiency and this has lead him to develop a language of building using brickwork that broke away from the traditional expressions of materiality and its structural use, of heaviness and solidity. He has created structures in masonry that are slender and light and often appearing to defy gravity. He has also developed many new construction techniques, such as pre-stressing that are in themselves significant and ahead of contemporaneous developments elsewhere.

The work of Dieste shares similar traditions with others, perhaps better known engineers and builders, such as Torroja, Candela, Nervi and Freyssinet, remembered for their striking structures in reinforced concrete, a modern material, so evidently of the 20th century. The contribution of Dieste to contemporary architecture certainly matches their achievements. Dieste was an innovator of both form and materials and, in the re-discovery and re-interpretation of a traditional material in a modern, technically sophisticated manner that was appropriate to the needs and resources of Uruguay he was, by custom and practice, an environmentalist.

This book is not intended to be a biography, but simply an introduction to the work of Dieste. It is provides a broad overview of his work, it considers the structural forms he has developed and describes selected key projects. The penultimate chapter describes a series of projects where his ideas and techniques have been adapted and applied in Europe. There is more to be said about Dieste than is covered in this book. During his career he built many other projects not described here and he has made a significant contribution to mathematical analysis of shell structures. It is hoped that this book will stimulate further research into his life and work.

On the 19th July 2000, while this text was in the final stages of publication, sadly Eladio Dieste died after a sustained period of illness. I will always remember my last meeting with him and the sparkle and glint in his eyes as this modest and gentle man from Uruguay shared enthusiastically his ideas and described his projects, even though he was quite ill and confined to bed.

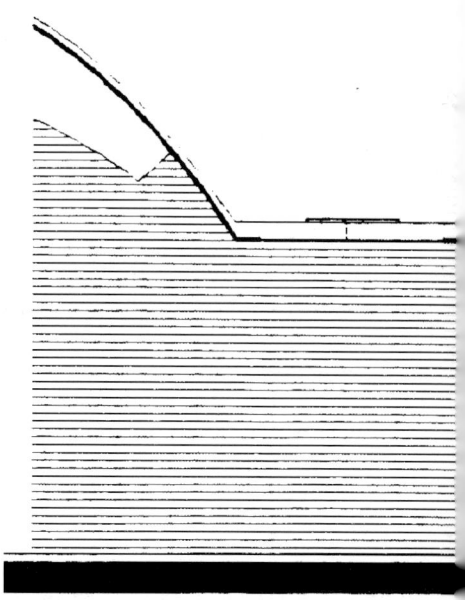

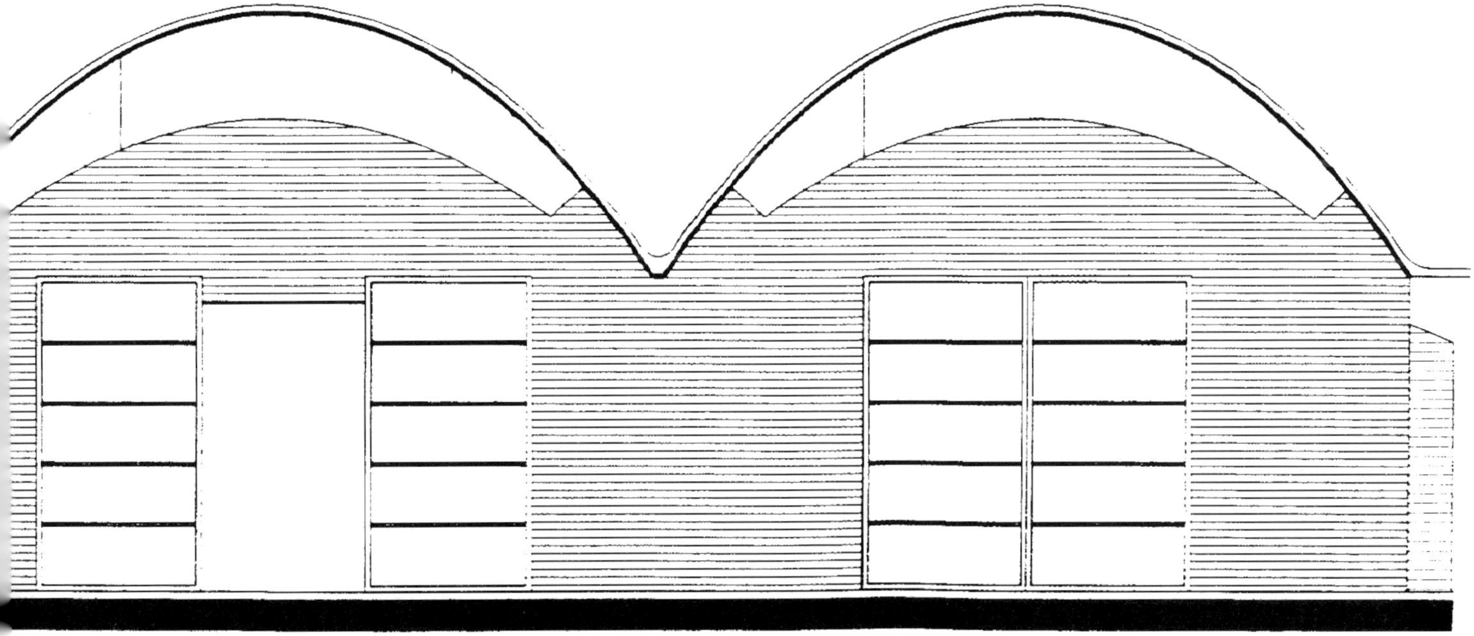

Contents

One	Introduction	11
	Dieste y Montañez SA	17
	Ideas and writing	21
	Innovation in brick structures	23
Two	The Master Builder - free-standing vaults	27
Three	Only the essential - Gaussian vaults	45
	Shallow vaulted roofs	47
	Horizontal silos	58
Four	The reluctant architect	65
Five	In the footsteps of tradition	81
Six	Towers	95
Seven	Typological variations	105
	Montevideo Shopping	106
	Casa Dieste	112
	Parador Ayui	117
	Nuestra Señora de Lourdes	118
Eight	Crossing the Atlantic	121
	The parish church of San Juan de Avila, Alcalá de Henares	124
	Nuestra Madre del Rosario	131
	La Sagrada Familia, Torrejon de Ardoz	135
	Camino de Los Estudiantes	135
	Insitut für Tragwerksenturf und Bauweisenforschung	139
Nine	Conclusion	142
Endnotes		145
Selected List of Works		152
Bibliography		154
Index		156

Chapter One
Introduction

Chapter One
Introduction

Studying the work of the Uruguayan engineer, Eladio Dieste, is like dismantling the pieces of a complicated and wonderfully crafted puzzle, each piece precisely made and carefully considered in relation to one another. The finished puzzle is both beautiful and intriguing; but it is only by completing the puzzle that the skill of the creator is revealed. In Dieste's work there are layers of architectural form, of structure, of construction, of detail and material, each one intact in its own right, each one able to justify itself within its own terms. On reassembly of the layers a new understanding and clarity appears, and one is left with a sense of the creative genius. On visiting his buildings one is immediately struck by the light and the form of the surfaces; subsequently, on reflection, one starts to experience the structure and its relationship to the form. With detailed consideration and investigation one sees the clarity, simplicity and expression of construction and, by comparison with other forms of construction, one sees the intelligent use of materials (that such sophisticated design and technically ambitious forms could be produced economically with brick).

The experience of the breadth of his work is also the discovery of considerable depth and application. Most people who know anything of Dieste will have seen the church at Atlantida, his most famous building that was produced early in his adventures with brickwork. However behind this there is an extensive compilation of buildings and projects that follow a trail through contemporary building typologies; factories, warehouses, sports halls, shopping centres, houses, water towers, petrol stations and concert halls, all sharing similar characteristics, but more significantly representing the exploration of a philosophy based on truth in form and materials, driven by a conviction that industry, innovation and technology must serve humanity rather than enslave it. Dieste has developed a unique language of building, characterised by surface forms in structural brick with echoes of a traditional past, but nevertheless based on contemporary structural theory and construction methods.

Dieste is a man of sincere conviction, a devout Catholic but not in a pedantic or ritualistic way, always striving for expression of the human condition. He has great faith in 'La gente sencilla' the 'modest people' the ordinary people who use buildings, who make buildings, who give life to space in an innocent and everyday way. Clearly, he feels a responsibility towards the people who use his buildings. In a rather respectful manner his churches illustrate this sensitivity and subvert the established hierarchy of the clergy in favour of the laity, a return to the essence of the mass. As he has said:

'The Eucharist at mass is the sacrifice of our faith, but it is also a meal around a table'.

An element of subversion exists throughout all his work. In fact, his work could not exist without subversion: he challenges existing ideas and preconceptions on the use of technology and the import of technology; he clears the deck of the accepted, empirical conventions of structural engineering and construction and starts from the essentials.

When this is understood his seemingly complicated buildings acquire a clarity and a natural simplicity. His ideas have particular relevance today, as a philosophy of building, born in a developing country in Latin America during the middle of the last century, that is now, at the turn of the millennium being imitated and copied in the developed world.[1] His thinking has evolved in part due to the particular circumstances of his native country, which has a strong cultural and intellectual base, centred almost entirely in its capital but lacking the natural resources of the larger Latin American countries that have expanded rapidly into major manufacturing economies. The country therefore has had to find alternative and simpler ways to develop, based primarily on its agriculture and climate.

Uruguay is a small country, around one third the size of Spain, with a population of just over three million. It is located between the two largest and most powerful economies in South America, Argentina and Brazil. The country was created to act as a buffer between the two. Although Uruguay was discovered centuries earlier, extensive colonisation by Spain[2] did not start until the eighteenth century in an effort to limit Portuguese expansion of Brazil towards Argentina. The country eventually gained

Fig. 1.1. Portrait of Dieste.

independence from Spain in 1830. Uruguay is a flat, fertile land of plains and gentle hills with few natural mineral resources, the main sources of income are from agriculture and associated meat, leather and dairy processing. Livestock outnumbers the human population by ten to one. The country relies heavily on its two neighbours for both imports and exports.[3] The population is largely Spanish and Italian in origin and has the highest literacy rate and one of the highest standards of living in the continent. The capital city Montevideo, on the estuary of the River Plate lying opposite Buenos Aires, is the only large city and hosts nearly 50% of the total population of the country.[4] The remaining population is spread over 20 much smaller towns and in the countryside.

Chapter One
Introduction

Eladio Dieste Saint Martin (Fig. 1.1) was born on 10 December 1917 in the town of Artigas,[5] the furthermost town from the capital Montevideo, near Uruguay's northern border with Brazil. The Dieste family originated from the town of Rianjo, near La Corunna in northern Spain. Dieste grew up in a middle class liberal environment.[6] His father, also Eladio (1880–1972), taught history. Dieste formed a close relationship with his uncle Rafael Dieste (1899–1981), a well known intellectual and writer who returned to Argentina in 1939 exiled from Spain after the Civil War. Dieste shared a passion for poetry and language with Rafael. Another influence during his formative years was the painter Joaquin Torres-García (1874–1949) who had returned to Uruguay in 1934 from Catalonia. Torres-Garcia went on to establish El Taller García, a studio that would encourage collaboration among artists to produce sculpture, painting, architecture and crafts and which he used to promote his ideas on Universal Constructivism. *'What Torres- García envisioned as the new art for the Americas would encompass all expressions from architecture to the most humble utilitarian object.'*[7]

Torres-García became a close friend of the Dieste family.

In 1936, aged 18, Dieste went to Montevideo to study at the Faculty of Engineering at the University. Montevideo was flourishing at this time, becoming an important cultural and artistic city. In this first half of the century there was great investment in the fabric of the city. The profits of the country's agricultural exports were used to construct many civic squares, monuments, an extensive collection of Art deco buildings[8] and the expanding University. During this period, adding to the cosmopolitan and European atmosphere in the city, there was an immigration into the Montevideo from Europe of mostly Spanish and Jewish refugees. 'The generation of "45"', named after a group of film, art and literary critics, came to symbolise the maturing Uruguayan intellectual and cultural society.

The School of Engineering at the University of Montevideo, which produced its first graduates in 1934, grew against this background, confident it its own in regard to the city and country. The Engineering School placed emphasis on the fundamentals of mathematics and physics. Dieste left the University with a profound understanding of theoretical mechanics, inspired by his teacher Eduardo Garcia de Zunga to believe that the physical world could be described by the language of mathematics. Dieste quotes his teacher:

'The theoretical that fails in reality fails because it isn't theoretical enough.'[9]

Dieste would later demonstrate the confidence and desire to apply theoretical mechanics to structural and mechanical engineering problems.

'I am very passionate about the possibility of understanding reality by means of the physical-mathematical language.'[10]

Fig. 1.2. Casa Berlinghieri – Puna Ballena.

Fig. 1.3. Early Gaussian vault under construction.

Fig. 1.4. Gaussian vault.

Chapter One
Introduction

Fig. 1.5. Load test on a segment of vault.

Fig. 1.6. Early brick vault.

A characteristic of Dieste's approach to design is this ability to see structures in both a physical and mathematical sense and this to some extent sets him apart from other engineers (such as Heinz Isler and Frei Otto) noted for their striking structural forms. They have used models to develop their forms. Dieste, however, rarely used models, relying more on the mathematical description of form.

After graduating he started work with Cristiani and Neilson, well known Norwegian contractors, on projects that included concrete shells. In 1944 Dieste married Elisabeth Friedham, a Jewish immigrant from Germany. Elisabeth converted to the Catholic faith and their family eventually grew to 12 children.[11]

In 1947 he collaborated with the Spanish Architect, Antonio Bonet. Bonet had moved to Argentina in the 1930s and later became one of the leading architects of the Rio Plata region.[12] Bonet was a founder member of the Austral group, most of the members of which, including himself, had worked for Le Corbusier in Europe. The manifesto of the group proposed a re-evaluation of the purely functional orthodoxy which they believed had been misunderstood and which failed to consider the human spirit:

'...functional architecture with all its aesthetic prejudices and infantile intransigencies, has arrived at intellectual and inhuman solutions.'[13]

Their views coincided with Dieste's as his concern for humanity is apparent in his works and his writings.

Dieste was engaged as structural engineer on the project, with Bonet, to design the Berlinghieri House in Punta Ballena, a holiday resort on the Atlantic coast. During one design meeting, Dieste without a great deal of forethought, suggested using a brick vault in place of the intended concrete shell. Bonet, with his modernist roots, naturally felt that a traditional brick vault would be too heavy and out of place. Dieste, however, was thinking more of a thin brick shell roof. Rather than an architectural education he had followed an engineering one with its preoccupation in theoretical mechanics, and it is unlikely that he would have been familiar with the traditional construction techniques that Bonet thought he had meant. Dieste approached the problem with a fresh, if somewhat naïve, set of eyes:

'I had no idea if clay had been used in similar structures nor had I come into contact with Catalan vaults.'[14]

Facing the Atlantic, the Berlinghieri House (Fig. 1.2) has an open elevation constructed from four parallel brick cylindrical shells. The largest, over the living room, spans 6 m. The house became one of Bonet's best known works and appears in a number of texts on contemporary Latin America architecture.

In 1948 Dieste started work with Viermond SA, a company that specialised in foundations and piles. Throughout this time, from 1945 to 1965, Dieste also held the post of assistant professor of mechanics at the University of Montevideo where he taught courses in dynamics and machine design. His early career combined the practical and the academic. He also lived in a social environment rich in art and culture, one perhaps conducive to the spirit of experimentation whereby he acquired the confidence to handle complex analytical concepts of the Gaussian vault and helped him to avoid the stereotypical preconceptions of the relationship between engineering and architecture (*Figs 1.3 and 1.4*).

Dieste y Montañez SA

In 1955 Dieste returned to the idea of the brick vault and with Eugenio Montañez, a friend and fellow graduate from the University, started the firm of Dieste y Montañez SA. The firm was progressive in developing reinforced brick construction techniques, exploring and rigorously testing their ideas of in a series of projects, each one a prototype for the next (*Fig. 1.5*). Through a series of increasingly ambitious projects and in the construction of progressively greater spans they became convinced and assured of its potential (*Figs 1.6 and 1.7*). From the initial ideas used in the Berlinghieri House, a new, rational and economic form of construction had been developed that suited Uruguay, having an appropriate architectural language and using inexpensive, indigenous materials.

The role of Montañez, a very able engineer in his own right, was very important to the firm and often his influence is understated. Although calm and considerate in his manner, Dieste was held in awe by many of his staff and Montañez was the only one within the firm who felt comfortable in challenging his ideas, assuming the role of Devil's advocate. He was a great force in getting the work done. The following anecdote helps illustrate their relationship.[15] During the construction of one of their more complex buildings, Dieste had assembled the workforce and had explained to them at length the ideas and methods to be used. After some time, Dieste turned to Montañez: 'I can see from their faces that they have understood'. To which Montañez replied, having also studied the faces of the workers, 'My friend they understand nothing'. And proceeded to explain to the workforce what they were to do. Montañez was the pragmatist to Dieste the idealist. If Dieste could convince

Fig. 1.7. Brick vault under construction.

Chapter One
Introduction

Fig. 1.8. Horizontal silo.

Montañez, then the idea would be pursued. Dieste provided leadership to the firm while Montañez took charge of commercial issues and project management, allowing Dieste time to create. It was Montañez who was largely responsible for the commercial success of the company. From the 1960s onwards the workload of the firm grew. At the age of 58 Montañez moved to Brazil on behalf of the firm and oversaw a number of very large projects for agricultural and industrial buildings. His business skills were used to consolidate and commercialise Dieste's structural innovations.

Typical of many successful partnerships they were very different characters, Montañez an outgoing agnostic to Dieste, the quiet and serious Catholic. They were necessary and complimentary to one another, a relationship that has allowed the

Fig. 1.9. Dieste on site with Vito Pacheco.

firm to design and construct over 1·5 million m² of buildings in Uruguay, Argentina and Brazil. The scale of some of the projects is impressive, some of the largest being the 150 000 m² of market space in Rio de Janeiro, the silos at Nueva Palmira with storage capacity of 94 000 m³ (*Fig. 1.8*) or the fruit growers pavilions at Porto Algere in Brazil. This last is a series of vaults 47 m wide and 280 m long. The firm was, in effect, a major design and build contractor that had developed its own innovative construction techniques.

The firm consists of a small staff of engineers and technicians based in their offices in Montevideo and a number of site-based staff. In a typical project a core of skilled site staff will organise and oversee the work. Additional labour is recruited locally on a project by project basis under the day-to-day management of one of his 'captains'.[16] Of these, most notable were Vittorio Vergalito, Edio Vito Pacheco and Alberto Hernandez employed by the firm for many years.

Vergalito emigrated to Uruguay from Campobasso in Italy in 1950 and worked for the firm for 38 years. A small, quiet man of considerable intensity from rural Italy — at that time a land that considered time in seasons and generations — he brought to Uruguay a sense of order of the world based on a sacred respect for work, and for the utilisation of materials and natural resources.

'To my father, Vittorio is the evidence of what he feels to be the truth… 'He is the person who best

Fig. 1.10. Piling plant designed by Dieste.

incarnates the philosophy on which his whole work is based.'[17]

Edio Vito Pacheco (*Fig. 1.9*) worked for the firm for 36 years and also came from a rural background. Different in personality to Vittorio, Pacheco was a natural leader with an expansive personality, the ideal interlocutor between Dieste and the labour force. He was:

'*…a leader who interpreted and conducted the cultural mixture when the rural worker is incorporated with the more urban and organised construction industry.*'[18]

Alberto Hernandez was with the firm for 30 years. He was physically very strong, with a gentle and quiet disposition that belied his appearance. An expression used by Dieste best describes his nature: 'the coarse hand that pats the small child'.

Chapter One
Introduction

Fig. 1.11. Dieste at work.

In these men Dieste saw noble workers from humble rural backgrounds who were not well educated formally but were intelligent and reliable and understood what Dieste was searching for and were able to respond to the respect and trust that was placed in them. They have been a considerable influence on Dieste, helping him to define his ideas on construction and the meaning of his work. Dieste has described the reactions he felt in the workforce on the completion of a complex but relatively minor construction.

'This man distinguished very well the difference between what is important because of its size and cost and something that touches us in the most profound way because it expresses to us the force that it produced but without feeling that force.'[19]

Together these men are the embodiment, what Dieste means by 'La gente sencilla'.

Either by accident or by a curious type of divine intervention, a company grew together at a time when there was great optimism and opportunity within Latin America. The combination of skills and mutual respect for each others contribution allowed Dieste's inventiveness to flourish and, more importantly, to be used.

The firm also specialised in foundations and produced a series of innovations, including the design of piling drilling rig for raked piles and foundations for large machinery (*Fig. 1.10*). Dieste has designed a novel form of foundation for turbines and cutting machines that avoids the brute force of mass concrete foundations by, instead, using carefully positioned raked piles at the corners of a thin reinforced concrete. The raked piles are positioned to resist both lateral and rotational movements of the foundation. He also designed a range of construction equipment for prestressing brickwork. This was largely out of necessity, due to the prohibitive cost of appropriate proprietary alternatives. In the fabrication of mechanical equipment he formed close relationships with the makers. He worked in close collaboration with Heinz Striew, the owner of Taller Francisco Suttener SA, a metal workshop.[20] In these workshops the moulds and supporting structures used in the brick vaults were designed and fabricated. A number of his designs for machines and construction methods were patented.

In the history of the firm there has only been two serious incidents in the application of the new technology of brick shell vaults. In 1978 a vault collapsed on removal of the props.

The collapse occured because prestressing wire for edge beams to resist the thrust of the vault had been omitted. No one was injured. The second incident in 1987 was more serious; a vault collapsed and several men were killed. The investigation and trial found that Dieste y Montañez were not responsible and that the fault had been in the supply of cement with a slower setting time than specified. By the time the verdict had been announced the firm had already rebuilt the vaults, and Dieste gave the money that the firm received in settlement to the families of the bereaved workers.

Montañez retired from the firm in 1991 due to ill health while Dieste continued to work into his late seventies, creating a new series of projects in Spain.[21] In 1995 Dieste retired from the firm because of ill health, although he continues to act as an advisor, with his staff visiting him at home. The firm is now managed by his son Eduardo, who maintains the values of his father. The technical director in charge of design is a fine engineer, Gonzalo Larrembebere Diaz.

Ideas and writings

Dieste (Fig. 1.11) has written a number of essays and books that embrace the technical and philosophical aspects of his work. Some of the essays were published in the book that accompanied a travelling exhibition of his work in 1996.[22] There he explains his views on various themes embracing architecture, structure and construction, technology and development. These ideas have shaped his attitude towards architectural design and inform, from an intellectual standpoint, his reinforced brick construction techniques.

The use of brick in developing countries is often seen as an 'appropriate' technology, as a low cost and low skill material. In this respect, the work of Dieste needs to be distinguished from that of other architects such as Hassan Fathy and Laurie Baker also noted for their use of indigenous and traditional local material. Fathy and Baker deal with countries that experience a different level of under-development from Uruguay. In an essay entitled *Arquitectura y Construcción*[23] Dieste challenges directly the predominance of the materials of the industrialised twentieth century architecture and in so doing applies his awareness of form and knowledge of structural mechanics, to create a refined and sophisticated contemporary material that pursues innovation in its structure and construction.

'The resistant virtues of the structures that we make depend on their form; it is through their form that they are stable and not because of an awkward accumulation of materials. There is nothing more noble and elegant from an intellectual viewpoint than this; resistance through form'.

Dieste requires that structures need to be justified, not only in pragmatic and economic terms, but they must also satisfy what Dieste calls 'cosmic economy'.

'Isn't it enough that we try to build structures that are simple, resistant and economical to construct, with what is normally understood for simplicity and economy. I do not hesitate in confirming that it is not enough. What is called simplicity is usually unjustified simplification and economy usually refers to money and its movement, economy in the financial sense. The things that we build must have something that we could call cosmic economy, that is to be in accord with the profound order of the world.' [24]

His ideas on cosmic economy relate directly to the relationship between architecture and construction.

'For architecture to be truly constructed the materials should not be used without a deep respect for their essence and consequently their possibilities. This is the only way that what we build will have the cosmic economy.'[25]

In this quotation he is almost paraphrasing Louis Khan[26]

'What do you want brick?' The brick replies 'I want to be an arch...'

However Khan's ideas of truth in brick construction do not characterise the use of brick in the twentieth century. Jonathon Ochshorn[27] has given us an account of the ambivalent attitude towards the use of brick throughout the twentieth century. Whether it is used as a structural material or as non-load-bearing cladding, whether it is appropriate in a modern architecture, how it is expressed, either deliberately as non load-bearing, or mimicking an underlying structural skeleton or simply as a chromatic and textural device are still potent questions. It has been used by and justified by both Modernists and Postmodernists.

Much of this discussion seems trivial when an understanding of the origin's of Dieste's concept is sought. Indeed the work of, arguably, the leading exponent of an architecture based on brick is not referred to in Ochshorn's article. Dieste asks brick a slightly different question from Khan's: 'What do you want to be in the twentieth century and how can I help?'. In answering the question, he extrapolates the catenary form of the arch, drawing on the great strength of brick in

Chapter One
Introduction

compression, and modifying the geometry to improve stability against buckling. The result produces structures that behave as traditional vaults but have a lightness that defies tradition and positions them firmly in the twentieth century.

In the same essay, *Arquitectura y Construcción*, Dieste discussed the predominance of the plane in post industrial revolution structures. The elemental nature of the steel frame, followed by the reinforced concrete frame meant that structures could be broken down and subdivided into two dimensional and, subsequently, into one dimensional elements, easily calculated using conditions of simple static equilibrium. In this way, then was established a mindset that persists today in many engineers and in their education; it is one that leads to simplification in design and analysis, almost to the extent of limiting creativity.[28] The fact that many of his works avoid the 'tyranny of the plane' makes them a curiosity for engineers for their 'apparent quirkiness', unorthodox and, hence, inefficient. His work has consistently received more attention for its architectural qualities; the profound logic in their form is often overlooked by the empirical rationality of conventional engineers conditioned to dissect structure into simple elements convenient for analysis. Computer-based methods of analysis such as the finite element method have given power to engineers to analyse virtually any form or shape, forms that have been developed without any reference to an inherent structural rationalism. Obvious examples include the Guggenheim Museum in Bilbao or some of the exhibits at the Millennium Dome in London. While providing freedom for the architect, the computer can also become a tool for further avoiding the rationality of structure, a tool that analyses structure indiscriminately without judgement. In the book, *Structural Engineering*, Bill Addis discussed the role of the calculation in the design of structure, airing a commonly held view about the predominance the numerical process:

'Nowadays young engineers are generally brought up without a good knowledge of precedent and believe that the mathematics of engineering science encapsulates all they need to know.'

Addis later considers the difficulty of 'putting theory into practice' and presents the contrasting opinion of engineers noted for their creativity — for example, Nervi, Candela and Torroja who believe that calculations are used merely as a confirmation of what is understood intuitively.

'The calculation of stresses can only serve to check and to correct the sizes of the structural members as conceived and proposed by the intuition of the designer.' (E. Torroja.)

With respect to his intuitive sense, Dieste is in accord with these engineers and his early structures, such as Atlantida, were conceived in a similar manner and calculated using simple approximations of the true structural behaviour. However, he has also pursued the mathematical analysis and calculation of his structures and presented his findings in two books[29] and a series of research papers.

A recurrent theme in his writings is the influence of technology and development on human need and its effect on 'modest people'. In two essays, *Ténica Y Subdesarollo* and *Arte, Pueblo, Tecnocracia*,[30] he is perhaps at his most political, considering attitudes towards development and the aims of development. Recognising, on the one hand, historic links with Europe and, on the other, the situation in his home country that must work with its own resources and its own heritage. He makes an important distinction between economic and human development. Economic development is often based on an interpretation of national statistics such as productive output per capita or average standard of living. It creates a pressure to produce and consume.

'This is the only way, through frenetic consumption that this productive machine will work.'

But what is the aim of development if it is not the betterment of mankind? The question has its roots in his own experiences, and his writings use many anecdotes as illustrations. As a young man touring Europe he saw at first hand some of the problems of industrialised economies, in particular housing 'with the barest of human comfort'. He believes that progress obtained at any cost is simply not worth it. Progress must include the greater fulfilment of mankind. While visiting an office in New York he was very impressed by the smoothness and speed of the elevator. He returned to the same office a few weeks later to find that the functioning lift was being replaced by one slightly faster and slightly smoother. The cost of the slight improvement in performance was very high and from Dieste's standpoint, a surprising and unnecessary waste of labour and resource. An example of production capacity and need driving consumption. In his view development should deal with science, health and art, fields of endeavour necessary for the fulfilment of human kind rather than the particular accumulation of wealth. He himself lives a modest life. In the third world, development often strives to echo that of the first world, no doubt conditioned by comparative economic statistics and conspicuous consumption. Technology is often imported at a considerable cost that is inappropriate to the context of the purchasing country. Dieste suggests an alternative starting point. Solutions to problems of development should come from within, from a clear understanding of the context and needs of the country and the application of first principles to the study of the problem. From here one can see the genetic components of his construction techniques. A sense of economy (in its broadest meaning), an understanding of form, abundant skill as an engineer and attention to the needs of the human spirit.

He believes that the ordinary people most often exploited or ignored by force of economic development need consideration. He believes further that these people can accept and appreciate beauty. Their needs extend beyond the purely utilitarian and indeed their cultural history shows that they have strong aesthetic impulses, expressed in a natural manner through their crafts and communities. To Dieste this issue is of great importance, if the lower classes cannot understand or accept art or beauty then, in Dieste's view, the controllers of development can feel justified in maintaining the low status and subsequent exploitation of these people. These thoughts are based on his experiences and his relationship with the people who build his buildings and who use them. They understand and contribute willingly to the effort that goes into building. Its aesthetic and moral character is that of the people themselves. The people who control development understand projects in notional terms of efficiency that allow the construction of factories and cities that are ugly and soulless. It is extravagant to consider aesthetics and utility. Factories become 'that infernal machine from which we escape at the weekend'.

Art becomes compartmentalised, controlled and in place, almost as a by-product of economic success but fails to recognise the needs of the simple people who find art in carefully crafted objects. Dieste's contemporary, Felix Candela,[31] had similar ideas.

'What is happening today is that painting is created by painters, music for musicians and architecture for architects, and no one thinks of the people.'

These same technocrats indeed are suspicious of attempts to include beauty in more functional buildings, a suspicion that increases the further down the economic ladder one descends. Factories should consciously express a lack of beauty on the theory that it will distract the worker rather than raise his spirits.

'Without meaning to they had created a new Moloch,[32] which was no less sinister than the original. This new Moloch was meant to efficiently crush the people for their own good and later they would receive good dividends.'

Dieste writes with some bitterness about a number of occasions where he tried, unsuccessfully, to convince businessmen of the value and importance of the consideration of form in these functional buildings. For example, a design he had prepared for a steel works was resisted by the clients. After some of argument, it became clear that they felt that a steel works should be ugly, there was no place for beauty in this type of building, the argument was not one of additional cost but of appropriateness. These ideas have developed during his career but their origins can be traced to the church at Atlantida, described in chapter four.

Innovation in brick structures

It is worth putting briefly into context Dieste's innovations in the structural use of brick. Further reference will be made throughout the book. Uruguay has a relatively under-developed timber industry and has to import steel and cement.

Chapter One
Introduction

Brick, however, is indigenous to the country.

The advantages of the reinforced brick construction are as follows:
- Brick is lighter in weight than concrete, reducing the cost of the supporting structure.
- In a brick vault, over 90% of the material is already hardened at the time of construction. The relatively dry brick absorbs moisture from the mortar in the joints, causing it to stiffen up rapidly and allowing moulds to be stripped sooner.
- Brickwork weathers and ages well.
- It has excellent environmental properties, its hygroscopic nature helps to control humidity.
- Brickwork uses less cement.
- Brickwork is easier to shape into double-curvature forms, than concrete.

The firm uses a variety of bricks, including hollow clay tiles. But Dieste's personal favourite is a handmade brick, manufactured in the open during the dry seasons and fired in simple kilns (*Fig. 1.12*). Each brick carries the mark of its maker, a building component intimately related to the human hands than formed it. In the book *Philosophy of Structure*, Torroja expresses well the character of brick:

'Bricks are considered to be the first materials created by human intelligence, from the four elements: earth, air, water and fire. This material, so close to the human spirit and need, being laboriously assembled and cast with skill...'

His buildings indicate a confidence in their builders to literally mould their traditional techniques into complex forms. At the time that Dieste was developing his brick structures, there was a great expansion of interest in the potential of structural masonry in the USA, Europe and Australia. A substantial international conference circuit had developed and a hierarchy of international experts, academics and professional engineers evolved. Much research concentrated on load-bearing, multi-storey buildings and the development of building codes and national standards. As the research field matured the attention turned to other applications including the use of reinforced brickwork. The first use of reinforced brick has been attributed to Marc Isambard Brunel in 1813, almost a century before reinforced concrete became established. However by the mid-sixties, reinforced brickwork was

Fig. 1.12. Handmade bricks.

used sparingly, to deal with crack control, for secondary elements such as lintels and to provide ductility to masonry buildings in earthquake zones. A review of the research literature on reinforced brickwork has been written by Sinha and Pedreschi.[33] Research efforts were directed to consider the use of masonry as an alternative to reinforced concrete and reinforced brick was used to imitate elements such as beams and columns. Dieste fundamentally disagrees with this approach:[34]

'It is possible to find rational uses for brick when combined with an adequate structure and suitable techniques which are not an imitation of what is done in concrete'.

The Brick Institute of America have provided a series of Technical notes describing the recent history and use of reinforced brickwork in the USA[35] and suggest that reinforced brick structures as suitable in the following circumstances:
- the structural medium is brick
- the surrounding area is brick
- the appearance of brick is desired.

Some large reinforced brick structures have been constructed, for example as St Hedwig's Church (1957) in St Louis, Missouri, where two 3·0 m deep, 19·5 m long brick beams were used to carry a precast concrete roof. However, these buildings tend to be exceptional and reinforced brick is used relatively infrequently, usually as a direct substitute for reinforced concrete.

All of this seemed to pass Dieste by. Even though his work was published in the architectural press in Europe[36] as early as 1961. The community of researchers and practitioners were largely unaware of his work, which was clearly far in advance of most that was going on at the time. In an updated article, Research and Innovation, in *Masonry International* Clayford T. Grimm, a leading USA masonry consultant, discussed the constraints to innovations in masonry, among which were fragmentation of both the research and construction sectors and the bureaucratic and lengthy procedures for the introduction of standards. Such issues have not hindered Dieste, because he has taken a step back from perceived wisdom and deliberately avoided empirical or quasi pragmatic 'shorts cuts' that frequently, in practice, stifle creativity. He has built with confidence, precisely because he has taken responsibility for the resolution of all the important issues of structure, construction and process including, as previously mentioned, the development of the underlying structural theory. A more cautious or less confident engineer seeking support from his peers might well have turned back.

Chapter Two
The Master Builder - free-standing vaults

Chapter Two
The Master Builder - free-standing vaults

Dieste has created two generic forms of vaulted reinforced brick structures: the free-standing barrel vault and the Gaussian vault. This chapter considers the barrel (*Fig. 2.1*), Dieste describes the barrel vault as *Cascaras autoportantes de directriz catenaria sin timpanos*, translated as 'free-standing catenary shells without tympanums'.[37] The barrel vault with single curvature was the starting point of his exploration of reinforced brick structures, initially with the Berlinghieri House and Antonio Bonet in 1947. One of the first projects of Dieste y Montañez was the competition they won in 1955 to design a warehouse for ANCAP. Their design was a reinforced brick vault 40 m long and spanning 8 m between a reinforced concrete frame. A series of further projects for storage and warehouses followed shortly afterwards, for Furgoni stores, also in 1955 and the El Pais newspaper in 1956. The family of free-standing vaults (*Fig. 2.2*) that followed include bus terminals, train sheds, service stations and, in 1995, the student walk-way at the University of Alcalá in Spain.

The barrel vault has its origins in the pre-scientific, craft tradition of construction. However, it is unlikely that the vaults that Dieste has designed would be recognised by the traditional master mason. The vaults are really rather the re-invention of a structural form with the benefit of contemporary analysis and interpretation than an evolutionary development of a traditional form of masonry construction.

The language and expression is of lightness and slenderness rather than mass and solidity. The associated supporting elements of the traditional construction, edge beams buttresses and tympanums, needed for overall stability, added to the weight of the construction and often obscured the extended aspect of the form, many of which are best appreciated internally:

'All the structures were primarily space-enclosing — in that sense inward-looking again usually with less, and sometimes with very much less, emphasis on the exterior.'[38]

Dieste's vaults have a very strong external expression, a voluptuousness deriving from bold curves and an exuberance in the sheer daring of the large cantilevers, paradoxically perhaps in a comparatively light structure, that clearly anchors them to the twentieth century of concrete shells and hyperbolic paraboloid roofs. One can sense an obvious delight in the demonstration of structural technique, a virtuoso performance by the maestro,[39] as the structure is boldly and precisely shaped in response to the internal forces. In most of the buildings the structure underneath is minimal in order to leave the roof floating, hovering above the floor below.

The arch and vault are some of the oldest of structural forms and replicate the natural phenomena of rock formations. Brick was the earliest prefabricated building component. Brick arches date back to the seventh century BC, a form evolved to utilise the inherent strengths and to avoid the inherent weaknesses of masonry. The

Fig. 2.1. Refresco del Norte.

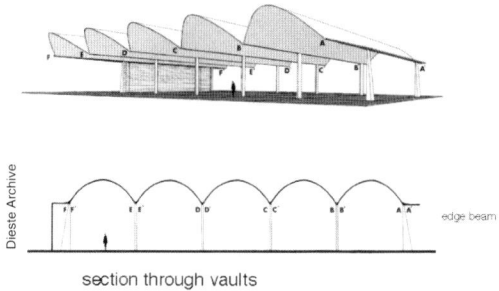

Fig. 2.2. Free-standing Barrel vaults.

Chapter Two
The Master Builder - free-standing vaults

arch, vault and dome were developed and used throughout the history of construction for the creation of large masonry structures that required power, ingenuity and ambition. The vault became used as much for its ability to create large, monumental spaces in both religious and secular buildings as its structural characteristics. Roland Mainstone in presenting the history of the building structures[40] stated that the vault was used 'by an equally non-structural desire to exploit aesthetically a new structural possibility'.

The arch continues to be an apparent structural and architectural form. In the nineteenth century, as iron and then steel became available and led to the development of the frame, the arch was still used in larger steel and iron structures, for example the 111 m span of the Galérie des Machines at the International Exhibition in Paris, 1889 or St Pancras railway station in London. Many long-span buildings of the late twentieth century use steel arches and vaults, probably more frequently now than the tensile, cable supported structures associated with High Tech architecture in the 1970s and 1980s, for example, the new Waterloo Rail Terminal,[41] the extension to the Imperial War Museum by Arup Associates and the Ponds Forge Sports Centre in Sheffield.

Similarly, as modern reinforced structural concrete[42] evolved, after steel, it too was used in arch and vault construction. Engineers such as Robert Maillart and Pier Luigi Nervi used concrete arches and vaults to build structures both beautiful and practical. Shell structures derived from the arch were used for their efficiency in materials and consequent lightness. As if echoing the words of Mainstone on the vault, contemporary architects employed the shell for its aesthetic rather than structural qualities, often to the detriment of the latter. Certainly, the most well known of these was the Sydney Opera House, by the Architect, Jørn Utzon. The difficulties in both designing and constructing a 'shell' which did not conform to the geometric and structural requirements for shells are well documented.[43]

During this time the use of masonry as a structural material for larger buildings declined. The mass of material, limitations in form, the generally slower rate of construction combined with a lack of understanding of the structural behaviour in terms of the developing language for steel and concrete, combined to relegate masonry to an in-fill or cladding for framed structures.

Before the Scientific Revolution[44] the understanding of masonry structures was largely empirical and experiential. Undesirable structural characteristics, such as cracking or movement, would be resolved during construction and avoided in subsequent buildings. Information and experience was passed on to younger apprentices, who themselves learned by observation and practice. Thus, there was a close relationship between the design and the construction of buildings, both roles filled by a single individual, who became known as the Master Builder. The Renaissance was responsible for the decline of the Master Builder[45] as the principal designers of monumental buildings, the role being taken over by artist–architects on the premise that the conception or design was more important than its execution.

The separate roles of designer and builder prevails today, However, there is a tradition of master builders that persists from the Industrial Revolution to the present day. These have usually been innovators and pioneers in the new structural materials, like Candela, Freyssinet and Maillart. The experience gained by trial and error has been replaced by scientific experimentation and analysis, which in turn has allowed for more rapid progress and broader dissemination of information. The work of Dieste is that of a Master Builder. There is a close relationship between design and construction. His innovation, typified in the barrel vault, is both in the conception and design of the form and also in the techniques of construction, developed and improved through successive projects.

It is helpful for an understanding of Dieste's work, firstly to consider briefly the mechanics of traditional masonry arches and vaults. The arch obtains its strength by maintaining a compressive force path from the point of application to the reaction points. The horizontal reactions at the support points, buttresses, are key to maintaining the stability of the arch. In a catenary arch,[46] the shape is determined by the self-weight of the arch

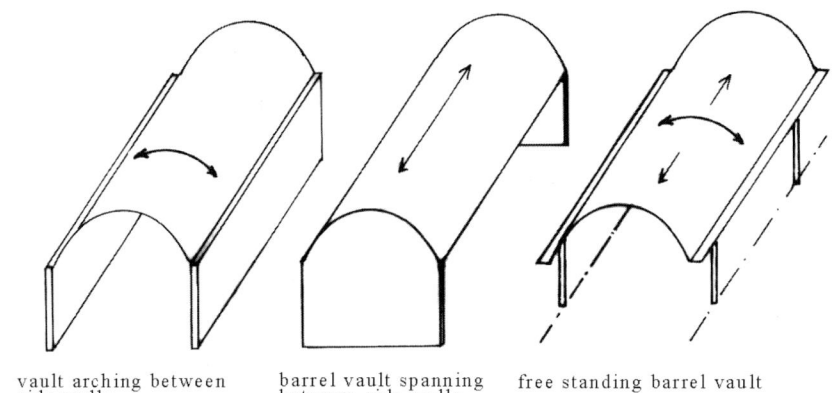

vault arching between side walls

barrel vault spanning between side walls

free standing barrel vault

Fig. 2.3. Various forms of vaults.

structural action occurs. The primary action of the structure is bending along the axis, the generatrix, inducing compressive stress across the top of the cross-section and tensile stresses along the lower sections. As the length of the vault increases relative to the cross-section, then bending will predominate and the geometry of the arch will not be maintained along the axis with the result that deformations[47] will occur across the section. Depending on the dimensions of the vault, load may be transferred by a combination, in varying proportions, of arch and beam action. When the longitudinal span exceeds three times the transverse span the dominant structural action is bending.[48]

Dieste's vaults have a much higher longitudinal to transverse span ratio; yet they are designed to allow both arch action and bending action to develop. A typical roof consists of a series of vaults connected at the valleys (*Fig. 2.4*). Between the valleys the vault takes the catenary shape, the ratio of transverse span to rise of the vault is typically between 2·5 and 4·0. The transverse span is relatively short and, consequently, the stresses in the masonry are low. The transverse section of the vaults acts as a deep beam that spans between columns underneath the valleys.

The complete roof can be considered as a series of Y-shaped beams, defined between the apex of two adjacent vaults. The beam is prestressed to avoid any flexural tensile stresses. The integrity of this construction allows long

itself to ensure uniform stresses in the cross-section. The catenary shape is identical to the natural sag of a cable under its own weight. Many arches are constructed that do not follow the catenary shape. For example, circular arches are easier to set out and construct, or arches may be subjected to different patterns of load which do not comply with the catenary shape. In either situation, the compressive stresses are no longer uniform across the section. However, the arch will still remain stable provided the line of thrust is maintained within the arch. If the line of thrust moves outside the cross-section, then instability of the arch will occur. In arches constructed using steel or reinforced concrete, there is additional flexural rigidity within the cross-section and uneven or unexpected loads can be sustained by a combination of arch action and flexural action.

The barrel vault in one form can be perceived as an extension of the arch, as a series of connected arches running along a line of supporting walls (*Fig. 2.3*), providing both horizontal and vertical reaction. Vaults are often combined in a series of parallel runs where the horizontal reactions of individual vaults are balanced internally requiring only buttressing at the end vaults. The support walls can be replaced by a series of columns carrying stiff edge beams upon which the vault sits. Suppose that the vault is made of a rigid material such as reinforced concrete providing continuity between the arch strips along its axis. As the spacing between the columns increases or the edge beam between the columns reduces in size, a different form of

Chapter Two
The Master Builder - free-standing vaults

spans between columns and long free cantilevers at the ends. In many vaults it is common to use tympanum or gables at the support points. According to Mario Salvadori:[49]

'Barrels should be supported on end walls (gables) or stiff arches so as to avoid unnecessary and costly buttresses or interfering tie rods.'

Dieste, however, uses neither gable walls nor tie rods and indeed views the gable with equal disdain to the tie rod.[50]

'One of the advantages of these vaults is that, as the theoretical calculation and the structures have proven, they can be built without the expensive and anti-aesthetic tympanums used in the classic free-standing vaults.'

Avoiding the use of tympanums is a further expression of Dieste's desire to express structure through form. He avoids the use of stiffening ribs and thickened edge beams to emphasise the surface structure. He is able to do this because the geometry he chooses to describe the vault is based on the catenary. Any other form, elliptical, circular or parabolic will try to deform to its corresponding funicular shape. In these circumstances, ribs, edge beams or gables are used to restrain the structure to the required shape. But provided the catenary arch is properly braced at its ends there is no tendency for it to deform transversely. In intermediate vaults the horizontal thrust of adjacent vaults react against each other, transferring the thrusts to the end vaults, which must be braced against horizontal

Fig. 2.5. Edge details - Refrescos del Norte.

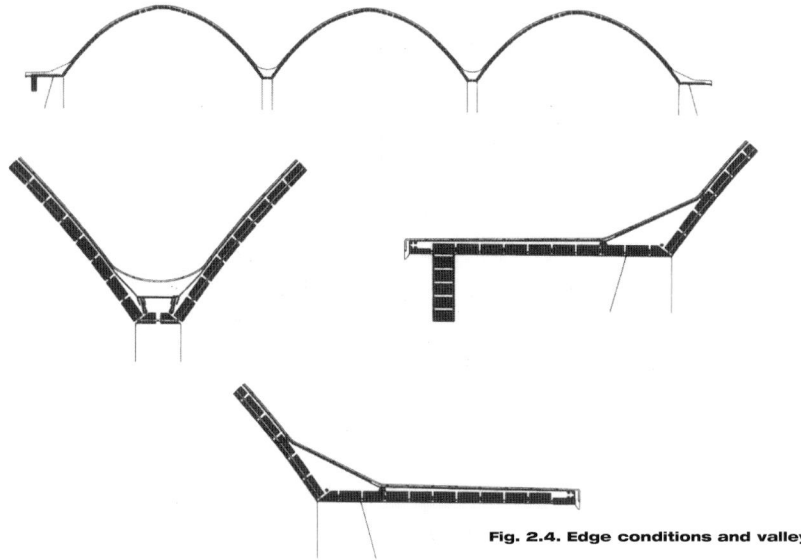

Fig. 2.4. Edge conditions and valley.

deformation. The sides of the end vaults, on the extreme generatrix, are stiffened against lateral deformation by forming a flat, horizontal edge to the shell. The edge slab is designed as a horizontal beam, collecting the thrusts from the vault and transferring it to stiff vertical buttresses. The action and relationship of this edge to the vault is critical both in performance and in the sequence of construction. The details of the edge beam vary depending upon the type of support. The edge beam may span directly onto end walls or be supported off columns placed inward from its edge (*Fig. 2.5*). In the latter case, the projecting edge of the beam is cantilevered from the vault itself. Thus the edge beam provides lateral support to the vault and the vault provides vertical support to the edge beam.

Construction of the vault starts with the edge beams, propped in the vertical position. The vaults themselves are constructed using a series of moulds. Each mould represents a segment of the vault, typically 5 m long. The moulds are quite simple, compared to the formwork for Gaussian vaults, being of single curvature and relatively short spans, constructed with timber on a lightweight steel frame. One mould is constructed for each separate vault, and construction of all the vaults progresses simultaneously. The moulds are used repeatedly. Great care is taken at the junction between adjacent segments of the vault to ensure that joints between the ceramic units are aligned and even; it is generally very difficult to spot the construction joint between two uses of the mould. The edge beam provides horizontal restraint for the vaults during construction. The valley beams are formed and propped until the vaults are completed. Removal of the propping is carried out in stages. Each stage causes various stresses and deformations to occur. As the mould is removed and re-erected for the next segment of the vault, load is transferred to the edge beam. On completion of the vault, the props to the edge beam are removed, transferring the weight of the edge beam to the vault or the side walls.

The next stage is the removal of the props to the valley beams, transferring the weight of the vault to the columns. Each beam will deform downwards between columns and at cantilevers. As the loads are transferred progressively to the final structure, the deformations of the edge beams require serious consideration. For the vaults to act as catenaries, it is essential that the edge beams be fixed in a horizontal position, since deformations in the edge beams will translate as deformations in the vault. Intuitively, given that the dimensions of the valley beams and edge beams are large relative to their span and that the loads are relatively low,[51] these deformations will be small. Nevertheless, bending stresses will be induced in the vault. The behaviour of the vaults is clearly complex and three dimensional; the longitudinal deformation of the valley beam will vary from zero over columns to maximum near the mid-span and at the ends of cantilevers. The transverse deformation of the vault will also vary as the valley beam deforms. The mathematics of this type of behaviour is rather difficult and translating this behaviour into constructible buildings marks the difference between conception and execution.

Dieste has developed the mathematical theory to calculate the additional stresses in the transverse section of the roof. The technique enables the vault to be reinforced to resist these secondary stresses. The theory behind the method, along with numerical examples, is described in the text by Dieste, *Cascaras autoportantes de directriz catenaria sin timpanos*[52] and includes design tables and numerical examples. In other hands, such technical expertise might well have been jealousy guarded rather than disseminated in considerable depth and detail.

The structural behaviour of these structures can now be resolved easily using computer techniques such as the finite element method. Subsequent computer analysis has verified his methods. However, the computer is normally only used to analyse a predetermined form rather than determine the form itself. Indeed, the computer has made it possible to analyse that which was previously unable to be analysed. In a curious way, it facilitates the structural design of complex forms that are conceptually non-structural. In the case of the barrel vaults, Dieste has developed a form that is pure in its concept and structure and it is difficult to imagine it being improved by computer. An underlying

Chapter Two
The Master Builder - free-standing vaults

Fig. 2.6. Prestressing vault.

of the construction. Instead, the roof is prestressed (*Fig. 2.6*) using tensioned steel reinforcement. The form of the vault and its economics makes it difficult to use proprietary or conventional prestressing techniques and anchorages.

Dieste has developed a number of different methods for prestressing to suit each of the different parts of the vault (*Fig. 2.7*). These methods are solutions to problems specific to the vaults. Pre-compression is applied wherever the predominant internal force within the vault is tension. The system adopted for prestressing over the supports consists of looped prestressing wires (*Fig. 2.8*). The loops are laid on the crown of the vault once the bricks have been installed and the

assumption in the theoretical analysis is that the structure behaves in an elastic manner.[53] Masonry is a brittle material, strong and sensibly elastic in compression but very weak when subjected to tension. The use of the catenary for the transverse section of the vault ensures a degree of pre-compression. In the longitudinal direction of the vault, large tensile stresses would occur in the upper part of the vault over the columns and in the lower parts of the vault in between columns and also laterally in the edge beams subjected to thrust from the vaults. These tensions could be resisted by adding more steel reinforcement; but the section would have to increase in thickness to accommodate the size of bars necessary, and cracking would still inevitably occur, compromising the weather tightness and durability

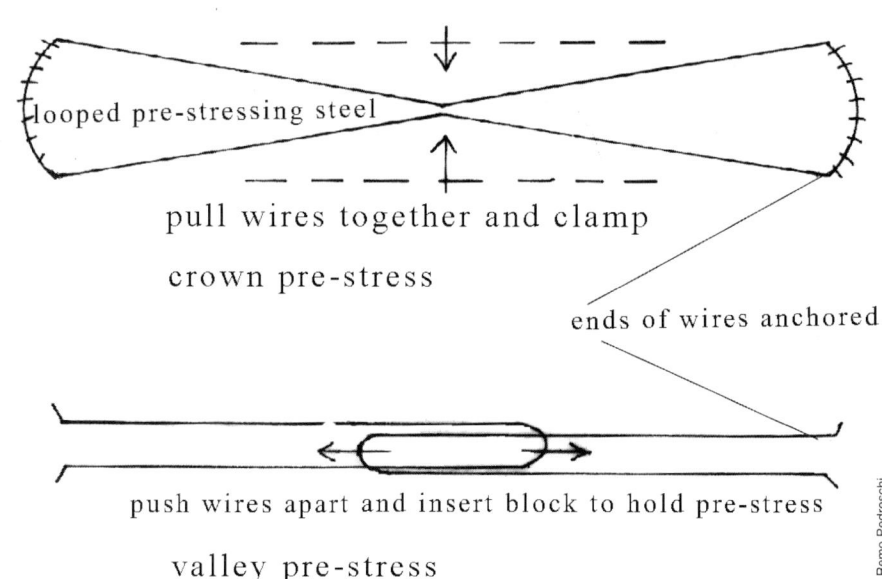

Fig. 2.7. Methods of prestressing.

Fig. 2.8. Looped prestressing tendons.

has a number of advantages over conventional prestressing:
- The large radius of the loop at the anchorage points ensures that the prestress is evenly distributed to the vault and avoids costly anchor plates and wedges.
- The prestressing equipment is much simpler and can stress up to eight or nine cables simultaneously.
- There is a simple geometric relationship between the width of the loop and its length. As the length increases then the width of the loop will also increase to provide a consistent extension.
- There is less loss of prestress force due to mechanical action. Using conventional barrels

Fig. 2.9. Vault under construction.

Fig. 2.10. Vault under construction.

reinforcement between the bricks grouted in place. Each end of the loop is embedded in reinforced anchorages, tied with steel rods to the vaults. The central part of the loop remains free and rests on top of the vault. The distance between the two sides of the loop is critical. Once the anchorages have sufficient strength, the loops are pinched together at the middle point, causing them to stretch and generate inward reactions at the anchor points, pre-compressing the vault. The wires are brought together using a simple screw jack developed by Dieste that straddles the loop. Once in their final position, the cables are held firmly using steel clamps. The top of the vault is then covered with a light concrete screed to cover and protect the cables (*Figs 2.9 and 2.10*).

Although the system is remarkably simple, it

Chapter Two
The Master Builder - free-standing vaults

Fig. 2.11. Agroindustry Massaro.

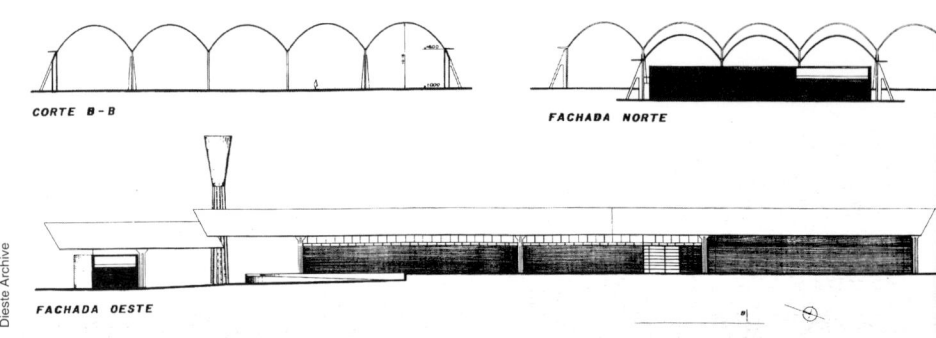

Fig. 2.12. Agroindustry Massaro - plan.

and wedges there is slippage during lock off, losing some prestress, this can be significant in shorter cables.

The second method of prestressing applies to the valley beam where the underside of the valley is subject to maximum tension. There is insufficient width to space the cables apart. The alternative uses two cables with overlapping loops. The ends of each loop are again anchored into concrete and tied into the vault. A specially developed jack is placed between the overlapping cables that pushes the ends of the loop apart, stretching the cables. Once the required extension has been reached, a steel block is placed between the two loops to maintain the separation of the cables. When the jack is released, the cables tighten onto the steel block, locking in the prestress force. Dieste developed the jack out of necessity as he was unable to obtain an appropriate jack in Uruguay. The jack is actually constructed using a standard lorry jack attached to a series of levers that converts the vertical action of the ram into a horizontal extension.

The warehouse for Agroindustry Massaro uses a series of long span barrels vaults to provide a large covered area for the storage of agricultural produce (*Fig. 2.11*). The construction and design was not straightforward, poor ground conditions and changes in brief that increased the size of the building lead to its present form The main part of the building comprises five parallel vaults, each 115 m long and 12·45 m wide (*Fig. 2.12*). The complete roof sits very disconcertingly directly on top of a series of infrequent concrete columns. The number of columns is kept to a minimum and clearly demonstrates the size and form of the roof as a single entity. The roof spans 35 m between columns and has a scalloped cantilever of 15 m at one end that projects over a second series of vaults covering the offices for the warehouse (*Fig. 2.13*). Both the cantilever and the second vault were incorporated at a late stage of the development of the project. The ground conditions needed careful piling and the large cantilever avoided the need for additional foundations. The secondary vault is supported off only one row of columns and cantilevers 15 m on either side for the same reasons. The secondary vaults are lower and slide under the main vault in a dynamic futuristic composition reminiscent of a shuttle leaving the giant mother ship.

Fig. 2.13. Agroindustry Massaro – overlapping vaults.

The vault itself is thin, only 100 mm, consisting of 75 mm of hollow brick topped with a 25 mm thick reinforced sand and cement render. The catenary shape of the roof permits this very high span-to-thickness ratio. Reinforcement is included within the joints between the bricks to counter the secondary bending stresses caused by flexure of the roof when the propping is removed. The cantilever roof is prestressed along the crown of the vaults using groups of 5 mm diameter wires in four staggered loops, each containing eight or nine wires (*Fig. 2.14*). When the wires are drawn together a pre-compression of 140 tonnes is applied to the vault. The initial separation for the outermost loop is 2·6 m. The valley is pre-compressed to a maximum of 96 tonnes using overlapping loops of 5 mm wire which are stretched to overlap by a maximum of 135 mm. The thrust of the vault is restrained by the side slabs that cantilever from the vault. These horizontal beams also cantilever 15 m in plan and are prestressed using displaced steel wires. The wires are initially straight and are pulled towards the edge of the slab where they are fixed with steel hoops cast into the slab. The beams themselves are fixed in the horizontal positions by buttressed concrete columns on piled foundations. Long cantilevers are generally disliked by engineers and loved by architects, disliked because they are difficult to control and it is simpler to prop the end, and loved for their boldness and apparent disassociation with

Chapter Two
The Master Builder - free-standing vaults

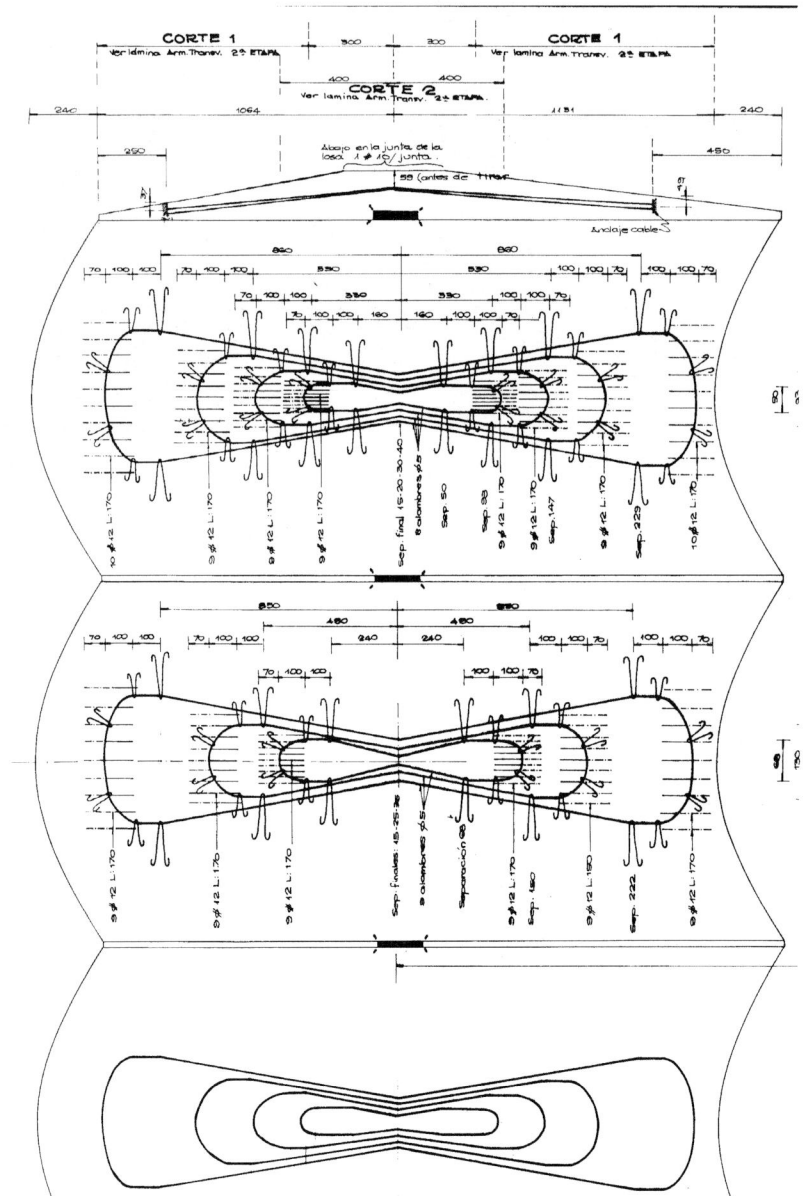

Fig. 2.14. Agroindustry Massaro - prestressing tendons in vaults.

structure. In Massaro, they are used confidently as an elegant solution to an unexpected problem that fits with the capability of the technology and the aspirations of the form.

The form of the vault, and its ability to cantilever long distances as used in Massaro, make it suitable for large canopy-type structures, providing shelter rather than enclosure. The principal form of national travel in Uruguay is the autobus. The country has an excellent system of cross-country buses with reserved seats, air-conditioning, televison and stewards serving refreshments. The bus terminus has the role and status of the railway station.[54] In Salto, situated in the north of the country, barrel vaults have been used in two principal bus stations. The bus terminal at Salto consists of a series of barrel

Fig. 2.15. Bus station - Salto.

vaults in double cantilever of 11·5 m either side of a central row of columns (*Figs 2.15 and 2.16*). The projecting edge slab is supported on a vertical edge beam in double cantilever, which is in turn supported from a projecting cantilever beam from the concrete column (*Fig. 2.17*). Structurally the edge beam is unnecessary, as the edge slab could have been designed to cantilever from the vault. The downstand of the beam was used to define the edge of the plan covered by the vault which runs parallel to the adjacent road. The edge beam, 23 m long in prestressed concrete, does not sit directly on the cantilever from the column, but is separated from it by a short steel stub. The vaults, lifted from the ground on as few vertical supports as are needed, attracts comparison with contemporary fabric structures. Both are form active and are used as roofs.

In fabric structures, however, the flexibility and form and the difficulty of bringing the roof edge to a straight line often create awkward junctions with the walls. The materials used in fabric structures are often advanced coated fabrics and it is essential to use a computer to define the geometry, analyse the forces and produce cutting patterns.

In buildings which are enclosed, Dieste often refuses to use the walls to support the roof, ignoring the potential to create a tympanum. Instead, the roof is suspended above the walls on concrete columns, with glazing in the space between. One example is the bottling plant for Fagar Cola (*Figs 2.18 and 2.19*). The roof

Chapter Two
The Master Builder - free-standing vaults

Fig. 2.17. Bus station at Salto - edge beam.

Fig. 2.16. Bus station - Salto.

overhangs the walls by 12·8 m on the end walls, which are glazed to the underside. As with many of his glazing details, the framing is kept to a minimum to preserve the visual integrity of the roof from inside to outside (*Fig. 2.20*). The glass is supported on steel bars, dowelled into the top of the walls. These details look simple, almost crude; however they are unlikely to leak. The roof overhang provides such protection that even the most torrential rainstorm is unlikely to wet the upper parts of the wall.[55] It is another example of cosmic economy; a roof that performs well structurally, that is also ideally shaped for shelter, with both functions symbolised powerfully in its form.

Fagar was constructed more recently than Massaro and a proprietary system of barrels and wedges was used. The 5 mm wires were replaced with 12·5 mm strand tensioned to a force of 12 tonnes.

At the beginning of this chapter Dieste was referred to as a Master Builder. There are a number of definitions for master, two of which seem most appropriate[56] are either 'a man having control or authority' or 'an artist of distinguished skill'. Both definitions coincide throughout his work. Dieste uses his artistic skill and control over technology to transform the mundane into the spectacular.

La Gaviota, (the seagull) a simple canopy over a petrol station becomes an object of wonder (*Fig. 2.21*). The complete roof, 17 m by 5·6 m, is constructed with reinforced brick and sits on a single central column. In structural terms it is a paradox, it takes its form from the catenary vault but transposes the form to become a curved cantilever slab, as if a single vault was cut along the ridge the two halves swapped and then re-joined at the eaves to produce a much less efficient structure. And yet! ... time and time again, when looking at his work, one sees signs of the apparently impractical, the victory of form over structure. But time and again, further consideration reveals pragmatic justification. A true vault would require two columns along each of the lower directrices. In the petrol station the most logical position for the vertical structure is in the central island where the pumps are situated as this provides the maximum space for manoeuvring vehicles, very important on this site so close to the road. Accepting this requirement,

Dieste Archive

Fig. 2.18. Fagar Cola.

Dieste Archive

Fig. 2.19. Fagar Cola.

Dieste Archive

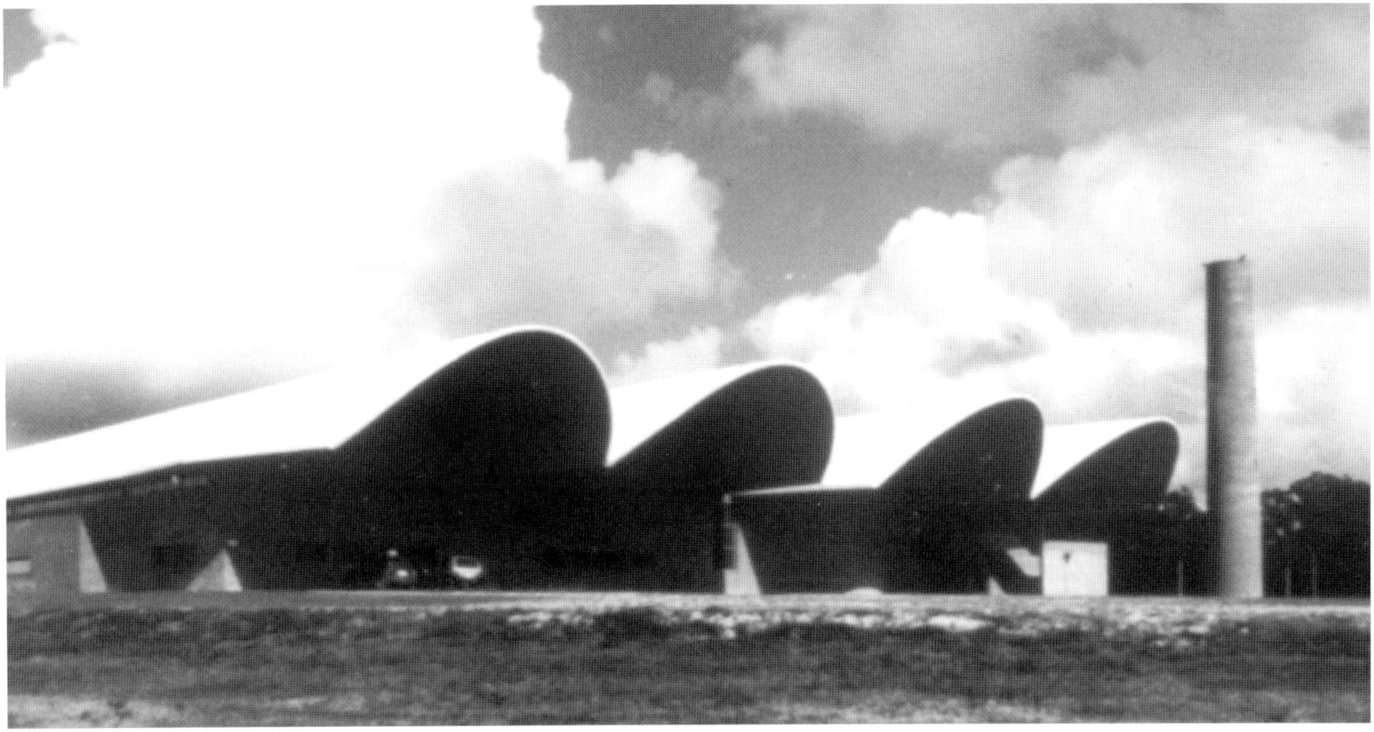

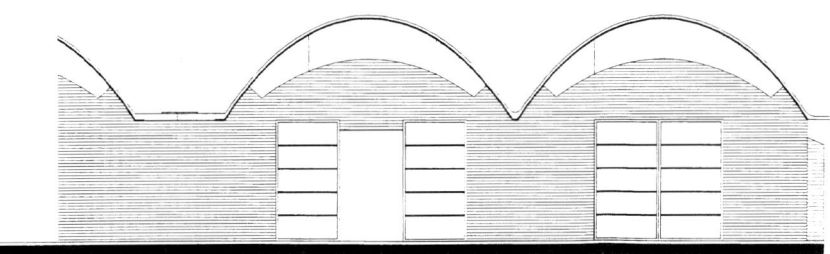

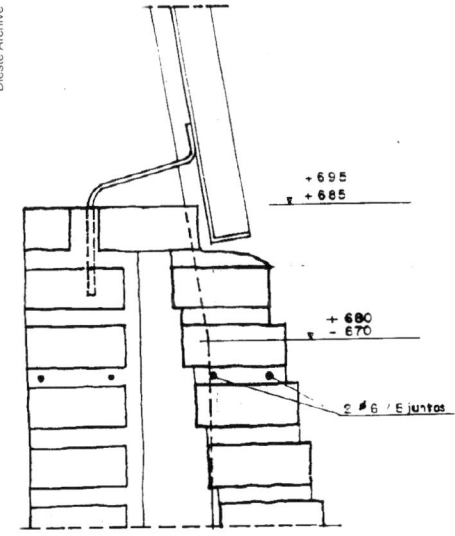

Fig. 2.20. Fagar Cola - window details.

Chapter Two
The Master Builder - free-standing vaults

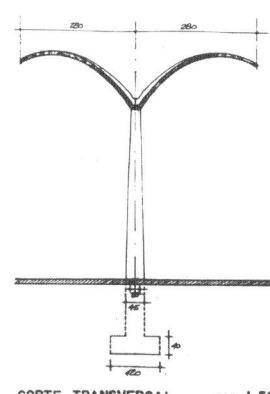

Fig. 2.22. The Seagull - section.

Fig. 2.21. The Seagull.

the roof has to cantilever, whatever form it takes.

The form of the Seagull uses the minimum vertical structure possible and thus provides the greatest access around the pumps. The gull-shaped cross-section creates sufficient structural depth to allow the roof to cantilever over eight metres on either side of the column (*Fig. 2.22*). The roof is prestressed along its long axis and reinforced across its section (*Figs 2.23 and 2.24*). The Seagull, constructed in 1976, has become an important landmark in Salto. In 1996, the area around the petrol station was scheduled for redevelopment, including the demolition of the Seagull. The mayor of Salto stepped in and declared that it must be preserved. A scheme was prepared by Dieste's son Antonio,[57] a successful engineer in his own right, to transport the structure across the city to a new location. A steel framework was assembled around the column to support the roof during removal of the column and transportation. The whole assembly was lifted onto a low loader truck and transported to a new site (*Fig. 2.25*). Once in place a new column was constructed underneath (*Fig. 2.26*).

With the development of the vault there are many innovations, and it is worthwhile putting these into context. In other countries, engineers were only starting to look at the potential of structural masonry in generally less sophisticated and ambitious structures.[58] Prestressing techniques for concrete, which since the initial developments of Eugené Freyssinet had become accepted as a technologically advanced form of construction, a specialised process that needed careful design and high quality materials, was still being developed. Freyssinet adopted the principal that concrete, as a manmade stone, is best suited to a state of compressive stress avoiding tension, a fact not lost to the Romans, nearly two thousand years earlier in the Pantheon. Reinforced concrete, however, is a technical solution to the lack of tensile strength, allowing concrete to be used in structural forms more suited to steel and developed initially for steel, namely framed structures. Prestressing is a purer form of construction in that it avoids tension rather than coping with it, allowing the full section of the structure to be effective.[59] As discussed earlier,[60] Dieste has an aversion to the frame as it encourages a reductionist approach to structure and ignores the potential of form resistant

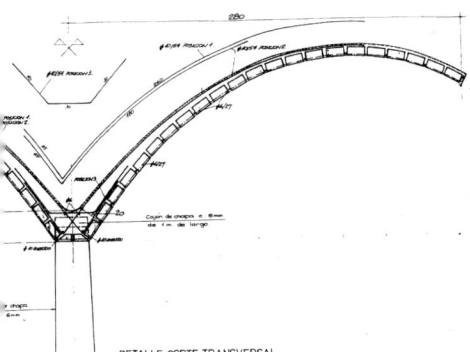

Fig. 2.23. The Seagull - reinforcement in vault.

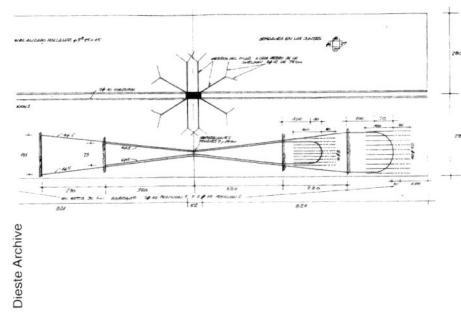

Fig. 2.24. The Seagull - prestressing tendons.

Fig. 2.25. The Seagull being transported.

structures. By prestressing Dieste is inducing the same stress condition that occur in the catenary vaults. In 1970 Ove Arup,[61] perhaps the most renowned engineer of the twentieth century, discussed the potential of prestressed concrete. He suggested that Freyssinet had invented a new structural material rather than improved reinforced concrete. He also believed that different techniques would evolve to eliminate the need for anchorages, looped tension bars for example:

'For instance, anchorages could be eliminated by providing the tendons in the form of continuous loops and jacking across a hole left in the structure and subsequently filled. But the manufacture of such loops of accurate length and strength must be mastered and the loops made available before even the economic validity of the 'idea' could be tried out.'

Dieste was clearly making innovations in materials, structural forms and construction techniques which were ahead of their time.

Dieste is a contemporary Master Builder, typified by the barrel vault where there is a close relationship between design and its execution. His innovation is to take an existing traditional structural material and apply scientific rationale to develop an appropriate form in lieu of experience and subsequently to refine and develop the construction methods to produce an established technique that is both efficient and economic.

Fig 2.26. The Seagull being lifted into position.

Chapter Three
Only the essential - Gaussian vaults

Chapter Three
Only the essential - Gaussian vaults

'*Only the essential finds a place in Dieste's work.*'[62]

It is one thing to achieve exciting architectural forms when budgets and expectations are high; it is another to achieve such forms in a truly pragmatic, functional and economic manner in buildings intended to be worked hard, as the farmer uses the tractor. More than any other of Dieste's structural solutions for buildings the Gaussian vault is perhaps the most coherent expression of his attitude to structural design. The Gaussian[63] vault (*Fig. 3.1*) has evolved out of the barrel vault, extending the use of the catenary to shallower and longer spanning vaults. Like the barrel vault, the Gaussian vault has found extensive application in a similar range of building types, large single-storey sheds, used as warehouses, gymnasia and workshops. Indeed, it is almost impossible to traverse Montevideo in any direction without encountering at least one of these structures. The curved geometry of these vaults (*Fig. 3.2*) has created an architectural form in the contemporary Hispanic tradition of Torroja and Candela.

The development of the vault has been for Dieste a relentless pursuit for 'cosmic economy'.[64] He has followed an inevitable path leading through the complexities of structural theory, construction processes and materials. As with many of his structures the outcome can appear improbable and anachronistic and appear counter to the perceived wisdom of contemporary structural engineering. But Dieste insists on the

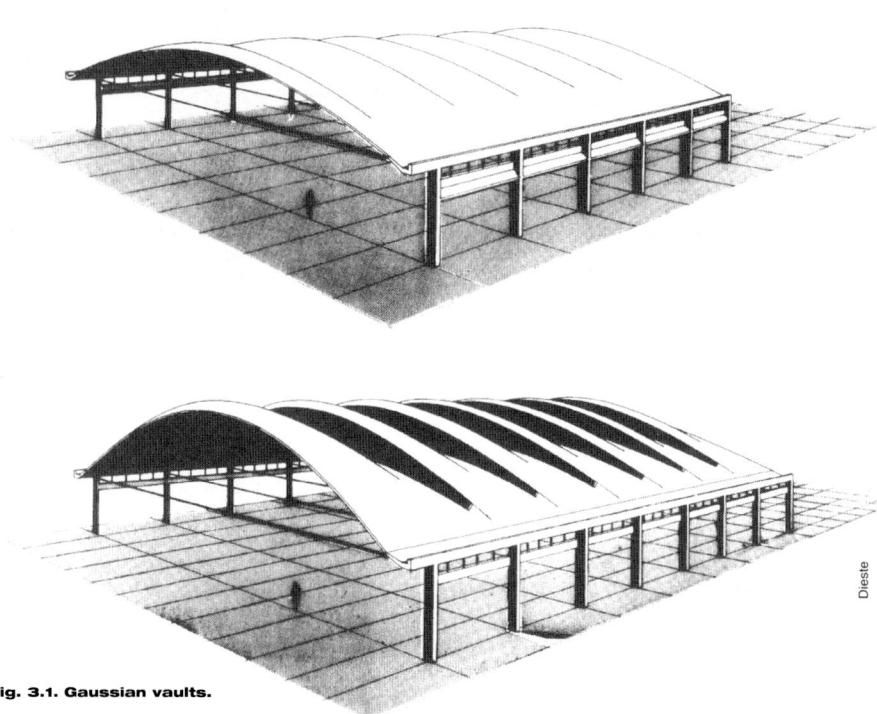

Fig. 3.1. Gaussian vaults.

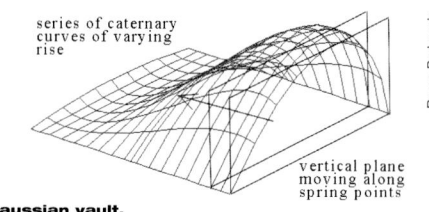

Fig. 3.2. Geometry of Gaussian vault.

Fig. 3.3. Vault under construction.

appropriateness of his way of working. Returning to the quotation in chapter one

'Isn't it enough if we try to build structures that are resistant, simple and economic to construct. With what is normally understood for simplicity and economy I do not hesitate in confirming that it is not enough. What is called simplicity is often unjustified simplification and economy usually refers to money and its movements.'

The search for cosmic economy has led to the formulation of a structure that 'resists through form', avoiding an empirical approach to design and ignoring rules of thumb, so loved by engineers and misinterpreted by architects, in short avoiding these convenient shortcuts from one known position to another known position. In the development of the vault, Dieste has started from a different position and followed a different route. The route has led logically and inevitably to the Gaussian vault. The inevitability of the route stems from a series of conditions, the use of bricks as an appropriate material in Uruguay, the use of the catenary and the desire to exploit surface forms. If the barrel vault was the product of a Master Builder, then the Gaussian vault is the product of a master structural engineer. Here the fabric is paired to its most minimal to create a rational and efficient structure.

There are two generic variations of the Gaussian vault that share similar geometric and structural roots but are used in markedly different applications: long span shallow vaulted roofs normally supported on a concrete frame or load-bearing walls and tall curved shells used in large horizontal storage silos.

Shallow vaulted roofs

These vaults take their form from the catenary arch, ensuring that all the forces will be transferred in direct compression (*Figs 3.3 and 3.4*). Although many of the arches are quite shallow, the stresses are nevertheless comparatively low in comparison with the strength of the masonry. The stresses are caused by self-weight. Increasing or decreasing the thickness does not change the compressive stress[65] in the vault. Considering only the axial compressive force, the thickness of the vaults can be minimal. In a vault of 100 m span and a rise

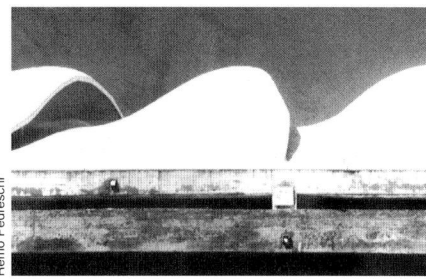

Fig. 3.4. Vault of Gymnasium at Durazno.

Chapter Three
Only the essential - Gaussian vaults

of only 10 m, the axial compressive stresses in the masonry due to self-weight will be less than 3Mpa, considerably less than the compressive strength of the brickwork. While very thin vaults can be postulated, assuming ideal structural conditions, the actual thickness is determined by practical considerations such as block size and thickness of finishes.

In *The way we build now* [66] rules of thumb are presented by Andrew Orton for masonry vaults that suggest a span-to-thickness ratio of 30–80 for practical applications. In Dieste's vaults, the thickness typically is 130 mm, and this has been used in vaults spanning up to 50 m — a span to thickness ratio of 384.[67] The main structural problem is not the axial stresses themselves but the thrust induced in such a slender structure, leading to a tendency to failure by buckling.[68] In single curvature vaults, barrel vaults for example, with large span-to-thickness ratios, such slender shells would tend to buckle under their own weight. Also of importance is the fact that the section is constructed using clay blocks with mortar joints, rather than close fitting stone voussoirs or monolithic concrete, but there is a greater sense of articulation and greater danger of deformation of vault. There are a number of possible, conventional options to deal with buckling. Either the compressive stress can be reduced or the resistance to buckling can be increased. The compressive stresses can be reduced by increasing the rise of the vaults. In single-curvature barrel vaults the rise is typically 25% of the span. In many buildings with longer spans, a high roof would be inappropriate for formal and practical reasons, (creating the wrong scale and adding unnecessary volume to the building). The resistance of the vault to buckling can also be improved by increasing the cross-section to make the vault stiffer.

The simplest and most obvious way is to make the vault thicker. This addition of extra weight and effort is a very cumbersome approach that contradicts Dieste's sense of cosmic economy. Additional material would not needed for the primary structural action, and the weight of the structure and loads on the supporting walls and horizontal thrusts will all increase requiring larger ties and anchorages, producing in the end a heavier, more cumbersome structure. A more refined solution involves the incorporation of arched ribs. Curved brick slabs would span between deeper arches. Conceptually, Dieste is uncomfortable with this solution: there would be a discontinuity in the flow of stresses between the arch and the rib, and structure would now be subdivided into simpler one and two dimensional elements, with a consequent loss in expression. More practically, the resulting structure would also be heavier, concentrating the forces at the anchor points of the arched ribs.[69]

Dieste has developed an alternative structure based on the use of form, rather than mass, to resist force. He has created an undulating surface which stiffens the vault in a similar manner to the ribs in profiled metal cladding. The resolution is in two parts. Firstly, the use of hollow clay blocks reduces the self-weight of the structure to approximately two thirds of an equivalent solid concrete vault, with a significant reduction in compressive stress and, hence, in buckling forces. Secondly, the shape of the structure is then manipulated to provide increased resistance to buckling without increasing the thickness.

The shape of the vault can be described using a family of catenary curves of varying rises. (*Fig. 3.2*)[70] The curves share a common springing point. Each catenary curve can be imagined as if contained within a vertical plane, the baseline of which sits on the sides of the building. The surface of the Gaussian vault can now be defined by moving this vertical plane along the line of the springing points. As the plane progresses along the axis of building, the rise of the catenary moves up and down, generating an undulating surface. The vault can be perceived as a series of interconnected vertical sections, each with gradually varying curvatures that together generate a warped surface. Along the central axis of the vaults, the roof has an undulating wave-like cross-section. At the apex of the vault these undulations are at a maximum, reducing progressively to zero along the springing lines. Therefore, the stiffness of the vault also varies from a maximum at the mid-point to a minimum at the supports, placing the stiffness where it is most needed, producing the minimum surface needed for stability. Reinforcement is incorporated between the bed joints to resist shears[71] between the different segments of the catenary, ensuring continuity in the vault. The vaults may be tied horizontally between the springing points using steel tendons connected to an in situ concrete framed structure or to load-bearing walls. In a number of buildings the ties have been eliminated for architectural reasons and alternative methods of resisting the thrust of the vault designed.

An apparently more straight forward solution is to use a cross-section of constant curvature, avoiding the warped surface for the vaults. However, doing so would create unsatisfactory conditions where the roof meets the wall. Dieste wanted to manipulate the geometry of the vault to meet the side wall more satisfactorily. In this respect there are significant differences between the Gaussian vault and the barrel vault.

Dieste exploits the form of the barrel vault as a structural beam allowing dramatic and striking overhangs to be used, freeing the vault from the edge of the building. He avoids filling the space between the intrados of the vault and the wall, the tympanum, to emphasise the strength and continuity of the roof over the walls. In the Gaussian vault the contact with the wall is a critical practical and visual detail, where the vault, anchorage and ties meet. Again he wishes to avoid creating a tympanum believing that it conflicts with the thinness of the vault, important in the expression of its form. He brings the vault

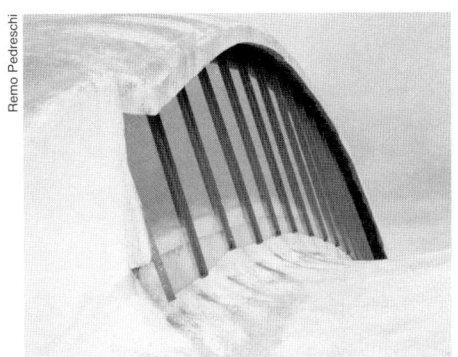

Fig. 3.5. Window detail - ADF Wool Warehouse

Fig. 3.6. Fire damaged vault - TEM Warehouse, Montevideo.

to a finish on horizontal eaves and in doing so delineates clearly between roof and wall. Ties, if needed, can be placed at this level, avoiding thrusts on the wall. The roof of the church in Atlantida is produced in this way (chapter 4).

A further development of the form is the incorporation of roof lights. The upper edge of one vault overlaps the lower edge of the next vault. A simple glass window is installed between the two vaults (*Fig. 3.5*). The underside of the vault acts a reflector for the light.

Although the vault is thin, the reinforcement adds considerable ductility to the construction. This was clearly demonstrated when there was a major fire at the TEM warehouse in Montevideo. A store of flammable plastic ignited and several cylinders of gas exploded. Two vaults collapsed as a consequence of the explosion and there was fire damage to the supporting frame (*Fig. 3.6*). A third vault lost a section 17 m long by 3 m wide. But the remainder of the vault stayed intact, held together by the steel reinforcement. The vault was subsequently repaired.

Dieste has developed a structure capable of spanning large distances, that is intuitively resistant to buckling purely by the form of its surface. Unlike other engineers noted for their catenary structures, such as Frei Otto, who made extensive use of models[72] to help understand geometry of fabric structures, Dieste did not use models. The models used by Otto were intended to define the basic catenary geometry, which Dieste could already do mathematically and were

Chapter Three
Only the essential - Gaussian vaults

therefore of limited use in anything other than a qualitative manner. Dieste's search was for a quantitative solution, based on the 'natural laws that govern matter'.[73] He needed to be able to calculate the stresses and ensure an adequate factor of safety against buckling of the vault, a catastrophic form of structural failure. This is by no means a straightforward mathematical problem; but having set forth on this particular path it is inevitable, by his assiduousness that he found its resolution.

As in so much of his work, he returned to first principles. Dieste has studied extensively the theories of elastic stability. He found that there had been little analytical treatment of the problem of buckling in double-curvature shells. And so he has developed the mathematical analysis of elastic instability to explain the behaviour of the vaults and condensed his findings into a simplified, approximate analytical procedure. This theoretical work is explained in his book *Pandeo de Laminas de Doble Curvatura*, (trans. Deflection in double-curvature vaults).[74] This text can stand on its own as an original and significant piece of research further demonstrating Dieste's profound understanding of theoretical mechanics. The procedures are used to check that the geometry of the vault is satisfactorily proportioned to ensure a minimum factor of safety of four against failure. Nowadays, catenary surface structures can be analysed accurately using computer-based methods such as finite element analysis[75] which Dieste y Montañez have used from time to time.

Had these techniques been available at the time, Dieste would probably have used them, as an aide to the development of the theory rather than as a substitution for the theory.

Brick masonry lends itself to the construction of double-curved vaults. Consider individual catenary lines. In principle, these could be made from accurately cut stone voussoirs[76] that do not require mortar, the stones being held by compression in the vault. However, when one considers the surface of the intrados of the vault and the manner in which it undulates, then the use of stone becomes substantially more complex. Clay blocks with mortar joints, however, allow the smooth articulation of the surface combined with a low-cost prefabricated unit.

Dieste's vaults are constructed on moveable formwork (Fig. 3.7, 3.8 and 3.9). The clay blocks are laid onto the formwork with the reinforcement placed between the joints. The joints are then grouted using a cement/sand mortar. A light steel mesh is placed on top of the vault and finished with a smooth sand/cement screed. The geometry of the formwork is complex and it is constructed using timber and steel. The design of the formwork is a major factor in the efficiency of the construction methods and it is fabricated as a single element for one complete vault, supported on a moveable steel framework which is raised to position the formwork at the correct height. In his design he replaced the more conventional hydraulic methods of lifting the formwork with a geared electro-mechanical

Fig. 3.7. Formwork for vault.

Dieste Archive

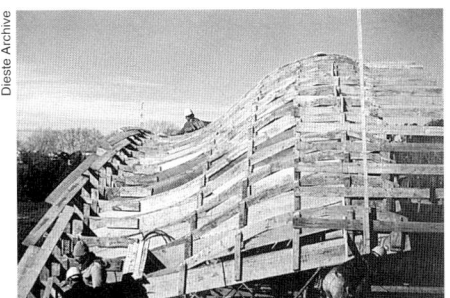

Fig. 3.8. Formwork for vault.

Dieste Archive

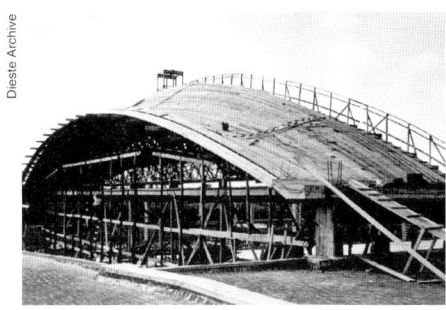

Fig. 3.9. Formwork for vault.

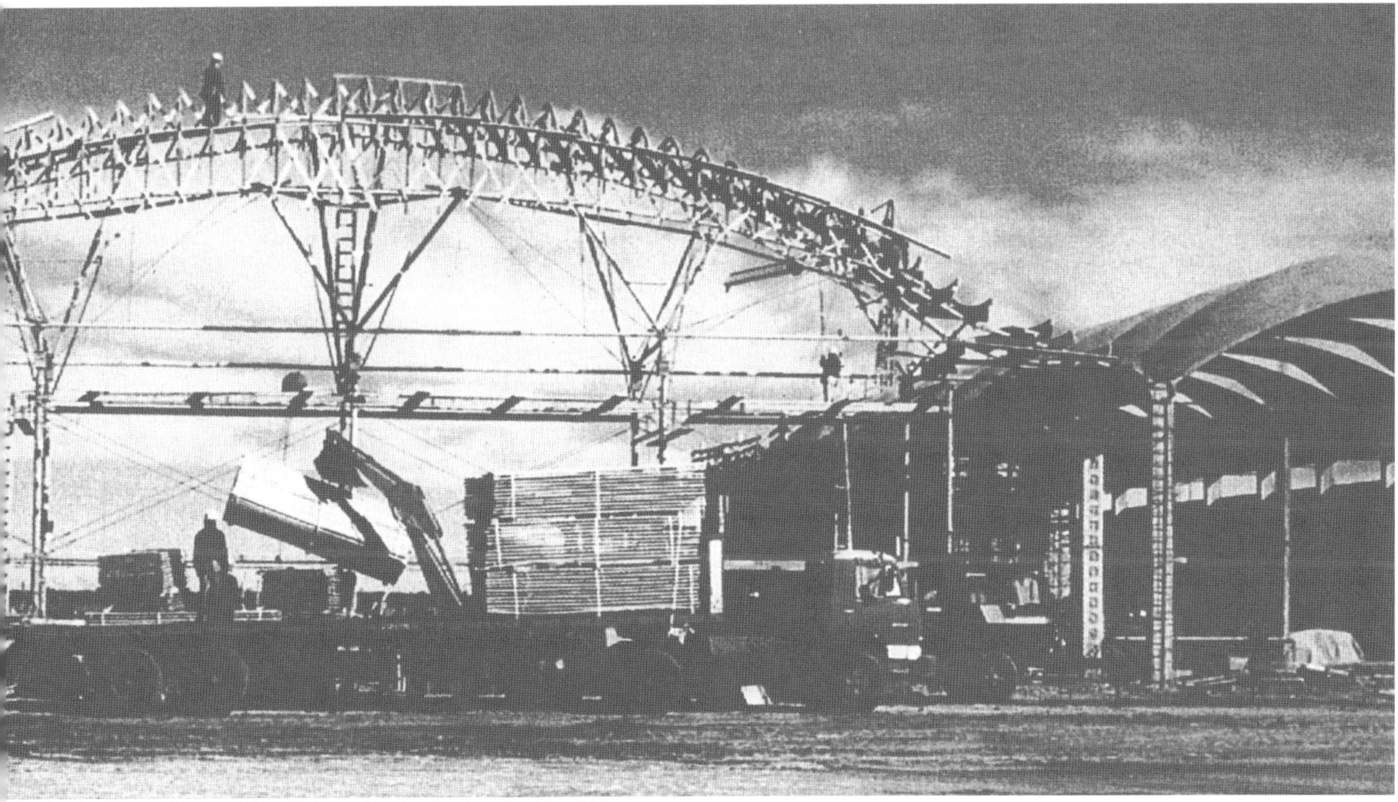

Chapter Three
Only the essential - Gaussian vaults

Fig. 3.10. Load test on vault - Dieste is standing.

A complete roof is formed by constructing a series of vaults using the formwork repeatedly. Although the formwork itself is expensive, it is reused 10 to 20 times. Using clay blocks, 90% of the material of the vault is already hardened as soon as they are laid on the formwork. The mortar, placed in the joints between the blocks, starts to stiffen as soon as it comes into contact with the blocks, as the water in the mortar is absorbed by the porous, hardened clay. The vaults gain rigidity more rapidly than concrete would. The formwork can usually be stripped after one day, two days if the weather is poor. In a typical vault the labour requirement is calculated as 1·6 hours for bricklayer and 1·4 hours for labourer per square metre of vault.

The process of stripping the vault is arguably the most onerous loading condition for the vault as the self-weight is transferred from the formwork to the vault. The vault is in fact subjected to an *ad hoc* load test when the strength and stiffness of the masonry are at their lowest. Dieste has also proved his design by impromptu load testing, using the labour force as applied load, uniformly distributed across the vault. In *Fig. 3.10* Dieste is standing at the apex of the vault.

Uruguay is largely dependent on its agricultural economy and many of the vaults are to be found in agricultural applications, a building typology normally at the lower end of the build quality spectrum of industrial building. It is normal to use local labour, usually agricultural workers, supervised by a small team of

system. The formwork, carrying a rather heavy load, has to remain in position for up to two to three days. With an hydraulic system the raised formwork needs to be locked in position mechanically otherwise any loss in hydraulic pressure would result in the formwork settling. Two systems are required, one for lifting and one for locking in position. The method that Dieste developed demonstrates his interest in solving problems in an harmonious and holistic fashion by settling both aspects simultaneously. An electric motor is attached to a wheel on each leg via a series of gears. The wheel rotates around a screw thread on the leg, raising or lowering the frame. When the motor is switched off, the formwork stays locked in position on the threads of its legs.

Fig. 3.11. ADF Wool Warehouse.

experienced employees from Dieste y Montañez. A typical building is the ADF wool warehouse situated in the Department of Canelones (Fig. 3.11).[77] A large building in three bays, used for the sorting, baling and storage of wool fleeces, it consists of three parallel vaults of 40, 25 and 25 m span. The vaults are constructed on concrete portal frames running along the springing points. Horizontal ties are anchored and prestressed between the portals (Fig. 3.12). The middle vault is 1 m higher than the other two vaults, to assist roof drainage and the installation of the horizontal ties. Large, crescent-shaped windows are formed between the top of the leading vault and the bottom of the next, in a similar manner to the sawtooth roof[78] configurations often used elsewhere. The glazing is constructed without frames, installed in recesses within the masonry vaults and accentuating the form created by the structure itself. It is as if Dieste does not want the window, only the form of the structure. Natural light floods into the building, reflecting off the intrados of the vault, creating a warm, top-lit internal environment. Even in dull, overcast days there is sufficient light to carry out the grading and sorting of the wool fleeces without artificial lighting. The building was completed in 1994 at a cost of $130 per square metre.

The Fruit Growers Market in Porto Alegre, Brazil, is one of the largest projects to use the Gaussian vault (Fig. 3.13). Dieste worked in collaboration with the architects, Fayet and Arajúo. The site contains a series of buildings for trading agricultural produce and was part of a

Fig. 3.12. ADF Wool Warehouse - tie details.

Chapter Three
Only the essential - Gaussian vaults

Fig. 3.13. Fruit Growers Market - Porto Alegre, Brazil.

Fig. 3.14. Growers Pavillion - Porto Alegre, Brazil.

Brazilian government initiative to encourage and increase agricultural production. The most impressive building is the Growers Pavilion, 290 m long with vaults spanning 47 m (*Fig. 3.14*). The form of the roof, although similar to the ADF wool warehouse, is more refined (*Fig. 3.15*). For architectural reasons, the use of horizontal ties was to be eliminated.[79] Instead, the thrust of the roof is absorbed at the edges by a horizontal slab, acting as beam inclined upwards slightly. The forces on the beam are transferred to the foundations by deep, vertical concrete buttresses. The projecting edges provide a continuous canopy over the open sides. The building is primarily a roof sitting on legs formed by the buttresses, clearly expressed on the entrances at both ends of the building (*Fig. 3.16*). Large windows are used to separate the vault from the entrances. The entrances themselves are solid, imperforate blocks jutting out from the pavilion adding further to the sense of separation between roof and wall.

To date, the largest span vault is the roof of the Julio Herrera and Obes warehouse at Montevideo Docks (*Fig. 3.17*). The span is approximately 50 m and the maximum rise is 6·5 m. The original warehouse was badly damaged by fire and a competition was organised to generate ideas for its successor. Most of the schemes presented recommended demolition and rebuilding. Dieste was attracted by the quality and modelling of the original walls and proposed

Dieste Archive

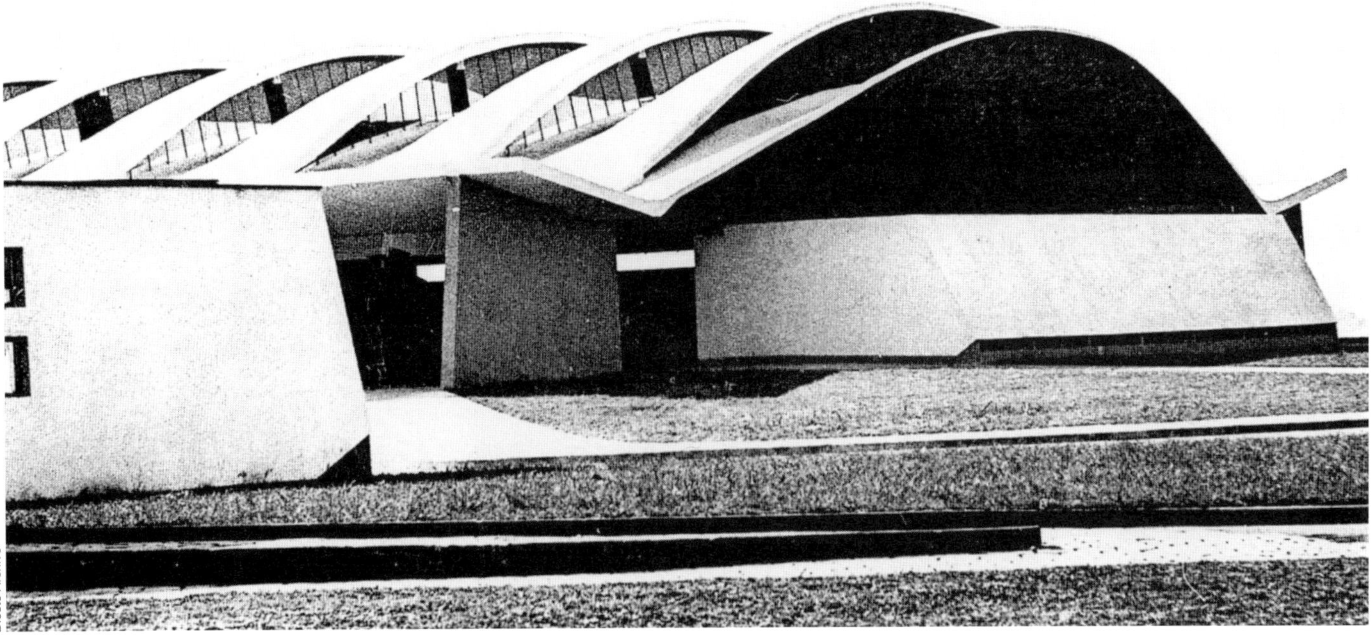

Fig. 3.15. Growers Pavillion - Porto Alegre, Brazil.

Fig. 3.17. Warehouse at Montevideo Docks.

Dieste Archive

Fig. 3.16. Section - Growers Pavillion.

Dieste Archive

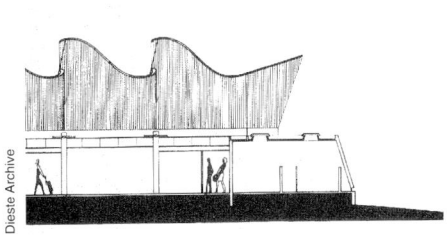

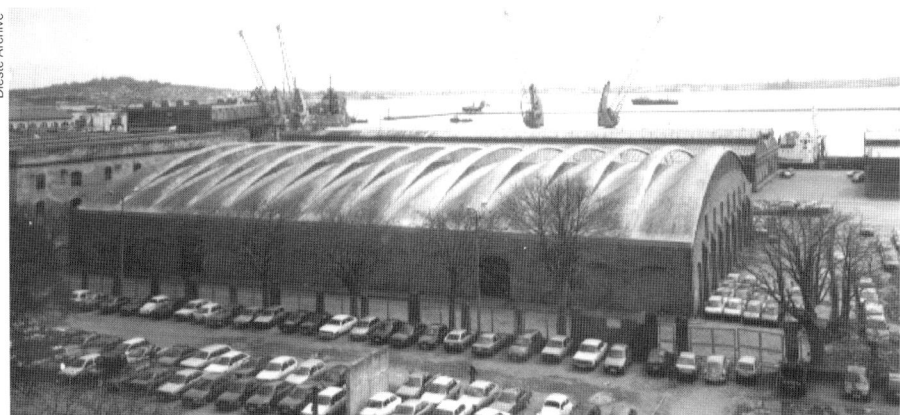

Chapter Three
Only the essential - Gaussian vaults

Fig. 3.18. Warehouse at Montevideo Docks.

a scheme to retain them and construct a new brick-vaulted roof. The roof form has a wave like quality, with a series of 14 discontinuous vaults appearing to tumble as they overlap (Fig. 3.18). The warehouse needed to be able to store anything from cars, to electrical goods, food and building materials. The vault spanned between the side walls, eliminating the need for any internal vertical structure and therefore increasing the flexibility of the space. Each 'wave' of the vault is 5·68 m wide (Figs 3.19 and 3.20). The windows at the front and back take their form directly from the leftover space between the lower edge of one vault and the upper edge of the other. The windows are again installed in grooves made during the construction of the vault. The interface between the roof and the side walls marks the transition between the old and new, and the point where both vertical and horizontal load transfer takes place. The shallow rise of the vault generates significant horizontal thrusts at the top of the wall, which it was never designed to withstand. These thrusts are restrained using a concrete edge beam running along the two side walls. Anchorage for the horizontal ties is incorporated in these edge beams (Fig. 3.21). The edge beam also marks the transition between traditional and modern construction techniques, the existing building and the new roof. The traditionally constructed warehouse is less accurate dimensionally than the prefabricated formwork and consequently the distance between

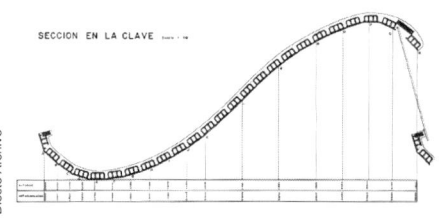

Fig. 3.20. Section through vault at apex.

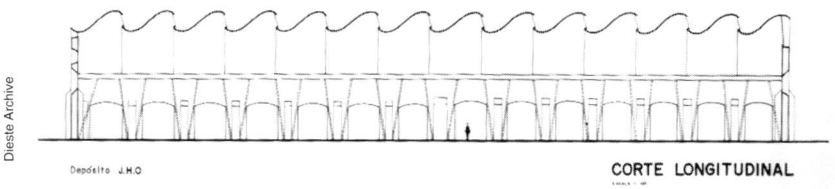

Fig. 3.19. Section through warehouse.

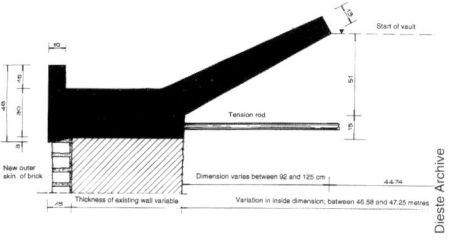

Fig. 3.21. Anchor of tie rod to side beam.

Fig. 3.22. Formwork for vault at Montevideo Docks.

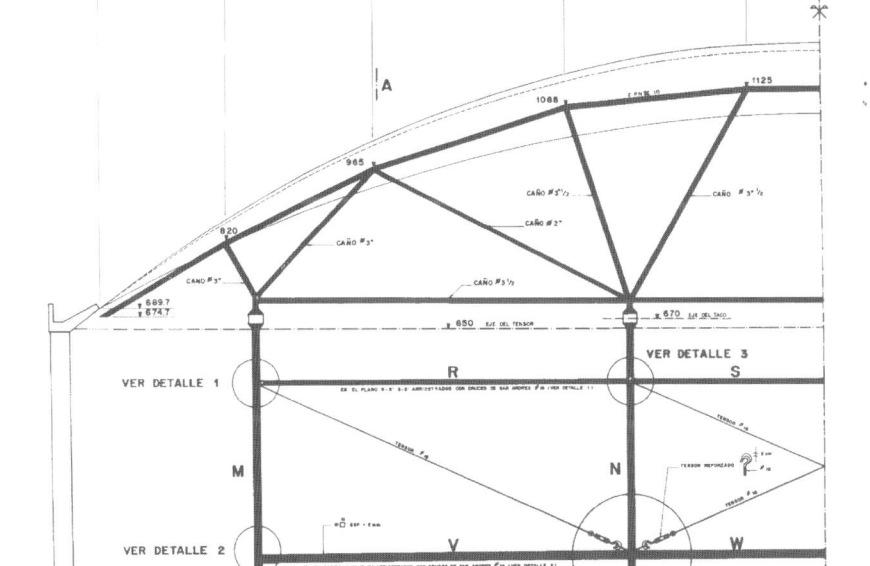

the two side walls varies along the length of the vault. The variation in the walls is accommodated using an edge beam that cantilevers inward from the top of the wall between 0·9 and 1·2 m, to ensure a consistent span for the prefabricated formwork. The formwork was constructed as a single large element to support one complete 'wave', 50 m in length (*Fig. 3.22*). It is a significant piece of structure in its own right. On the southern gable, the roof and wall are separated, highlighting further both the long span and the thinness of the vault (*Fig. 3.23*). The existing walls were re-clad using brick to follow the original modelling of the windows and buttresses and new brick arches constructed over the windows and openings.

Chapter Three
Only the essential - Gaussian vaults

Fig. 3.23. Warehouse at Montevideo Docks - gable wall.

Horizontal silos

The other form of Gaussian vault is the horizontal silo, a very large shed used for storing grain and other bulk materials. The structural principles and construction methods are essentially the same as for the shallow vaults, the form being generated by a family of catenary curves and the construction methods using prefabricated, reusable forms. At the same time there are a couple of important differences. The catenary is at the opposite end of the scale in terms of rise. Typically, the rise of a silo is one half of the span. There is also a major additional load produced by the thrust from the weight of the retained material. The stiffened curved shape is deepened[80] towards the base of the vault,

Fig. 3.24. Inside Solsiro silo.

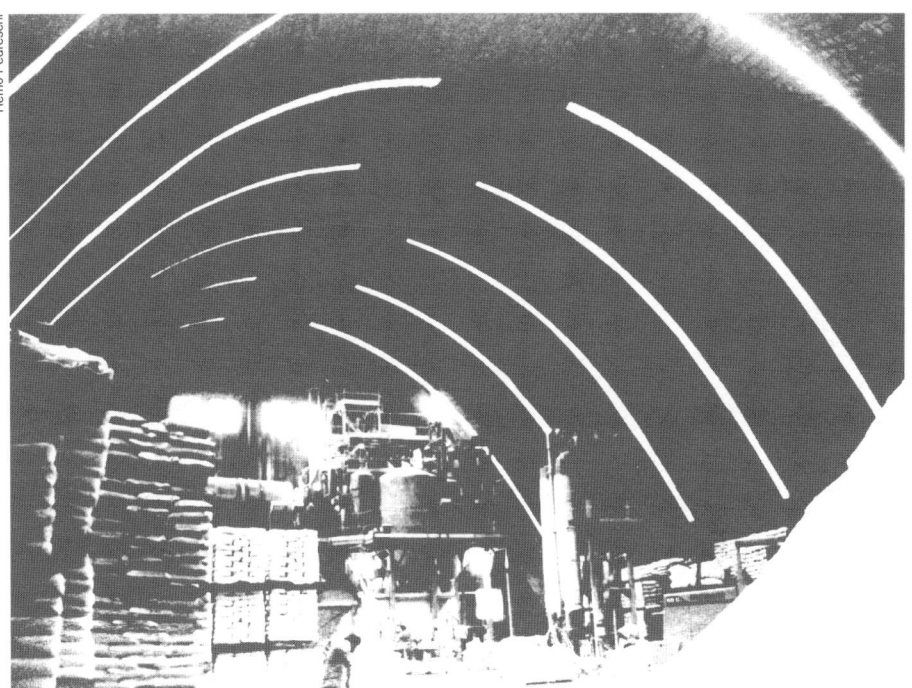

Fig. 3.25. Inside Solsiro silo.

increasing the bending stiffness to deal with these forces. The initial angle of the vault at the springing follows the natural slope[81] of the retained material thus minimising the thrust on the side walls. In developing a form based on the geometric properties of the warped surface Dieste has created a natural architecture like that of an elegantly proportioned suspension bridge. The visitor to these vaults is moved by both the form and the scale. *Fig. 3.24* shows a mountain of salt enclosed by a cavern of brick at the Solsiro silo. The internal surface of the vault is rendered using a cement-rich plaster, up to the maximum height of the retained material, to provide additional protection to the reinforcement (*Fig. 3.25*). Glazing is incorporated within the profile of the vault.

Fig. 3.26. Laying blocks - Silo at Nueva Palmira.

Chapter Three
Only the essential - Gaussian vaults

Fig. 3.27. Silo at Nueva Palmira.

Fig. 3.28. Formwork for silo.

One of the largest and the most recent is the silo at Nueva Palmira. The building is a major work of civil engineering. The silo is used to store soya grain. It is 142 m long, 23·7 m high and 45 m wide. The base of the silo is formed in a 7 m deep excavation. The side walls of the excavation are lined with reinforced brickwork. The horizontal thrust of the vault is resisted by embankments constructed using cement stabilised soil. The vault is connected to the embankment via a continuous reinforced concrete edge beam which undulates tracing the form of the vault. The height of the retained soya is 2·5 m above the top of the edge beam. The vault is constructed using 15 cm thick hollow clay blocks reinforced between the joints and covered with a sand/cement render (*Fig. 3.26*). The vault is

Fig. 3.29. Silo under construction, Nueva Palmira

finished with a layer of polyurethane insulation and finished with white acrylic paint (*Fig. 3.27*). The profile of the vault changes, getting deeper as it approaches the apex of the vault. A conveyor belt attached along the full length of the apex is used to fill the vault. The silo is discharged using a second conveyor belt constructed within a tunnel below the floor of the silo.

The steel framework that supports the formwork is a large structure, carefully designed and fabricated (*Fig. 3.28*). Large hot-rolled steel sections[82] are not generally available in South America and the structure is built up from small steel channels, braced and trussed to provide additional strength. The main sections of the framework (*Fig. 3.29*) describe the curvature of the vaults in a series of facets and consist of box

Chapter Three
Only the essential - Gaussian vaults

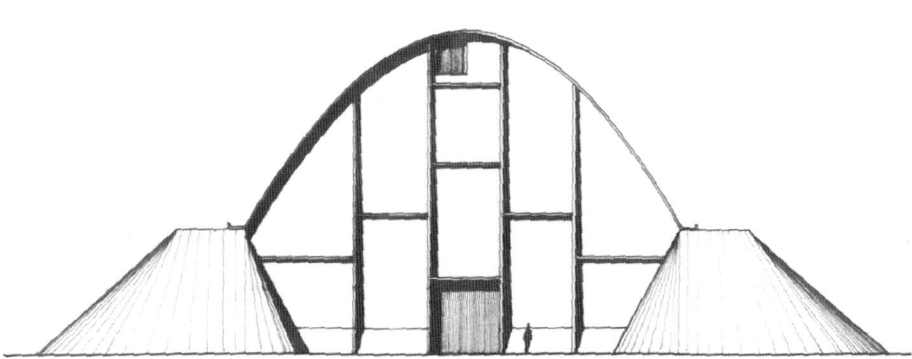

Fig. 3.30. End wall of silo.

sections made by welding two channels flange to flange and formed into trussed beams using steel rods and a vertical strut at mid-span. The framework is supported on five pairs of wheels and can be lowered and raised through a distance of 80 cm to strike the formwork. The geometry of the vault from springing to apex is described using 34 different timber profiles, lined with timber planks. The ends of the vault are constructed using inclined reinforced brick walls, constructed on top of the an enclosing abutment and spanning onto the edge of the vault (*Fig. 3.30*).

The project at Nueva Palmira was completed in April 1997. The horizontal silos are reminiscent of the airship hangars at Orly, France, designed by Eugène Freyssinet. The reinforced concrete hangars have a similar catenary shape, although the cross-section has a trapezoidal geometry.

In comparing the vaults with other structures, there is a dichotomy between the use of traditional material and contemporary analytical structural engineering. Traditional masonry structures typified by the Gothic endeavoured to create lightness by concentrating the forces into ribs, buttresses and groins, leaving much of the wall free from structure. The load-bearing elements themselves were heavy and indeed maintained their stability through dead weight. In other words, substantial piers were the price for open spaces. Dieste creates lightness throughout. He uses the form of the vault and the expression of its minimal thickness to create an impression of lightness, 'Leggero come un mattone', [83] Juan Martin Piaggio said. There is a clear visual expression of the thinness of the shells, the leading edges of the vaults appear fragile: there are no robust edge beams and just a suggestion that they might buckle. The essential difference between Dieste's use of masonry in these vaults and that of traditional masonry derive from his mathematical skills. He is able to analyse the structures carefully and acquire the confidence to extrapolate the vault from a thick heavy construction to thin light shell whose shape provides its stability.

The industrial building, typified by the large, single volume, single-storey shed has been described by Chris Wilkinson as the unwitting architecture of the twentieth century.

'There is a kind of architecture which is not formal, decorated or mannered, but which derives its aesthetic from a clear expression of its purpose and component parts where the demands of function and economy have led to simplicity of form and construction but where the basic requirements of enclosures and structure have been extended to create buildings of quality.'[84]

Wilkinson is not referring to Dieste's buildings in the above quotation although the definition does seem to fit the Gaussian vaults perfectly. He is of course referring to 'High Tech'. Like the Gaussian vault, many High Tech buildings such as the Renault Distribution Centre[85] or the Fleetguard factory[86] in France use funicular structures. The bold expression of the structure is central to their design. Masted tension structures can produce buildings that are both light in appearance and highly tectonic in their details. It is debatable, however, whether or not they are truly exemplars of function and economy in buildings rather than buildings stylised in a tectonic manner. Colin Davies expressed a sceptic view:

'But the technical disadvantages of exposed steel structures remain and no amount of justification (more economical foundations, column free interiors, increased flexibility) can dispel the conviction that the real reason for their popularity among High Tech Architects is that they convert ordinary factory sheds into colourful works of architecture.'[87]

Possibly the most rational and minimal of this genre is the Fleetguard factory in France, engineered by Peter Rice.[88] Nevertheless, such structures are essentially refined versions of the portal frame and employ masts and tension rods to reduce or eliminate bending in both beam and column. They are products of a process, mentioned earlier, that subdivides structures into two and then one-dimensional elements. Dieste, however, uses a material, perceived as traditional, in forms that are rational and avoids many of the problems of certain High Tech buildings — roof penetrations, the resolution of the roof wall interface[89] and the durability of exposed structures. They require a sophisticated understanding of materials, construction and structural mechanics and could be said in some ways to be more high technology than High Tech. In Dieste's own words:

'From the personal point of view all these things have been taken jointly into account: the validity and efficiency from the structural point of view, the efficiency from the economic point of view, the efficiency in the conditioning of the space enclosed and the aesthetic value, which for me, really, is a sort of ultimate summary, the ultimate synthetical value judgement of all this series of factors.'[90]

Chapter Four
The reluctant architect

Chapter Four
The reluctant architect

T he Church of Jesus Christ the Worker is the most widely known of Dieste's buildings (*Fig. 4.1*). Its completion in 1960 marked the point at which Dieste began to be recognised for his architectural skills.

'*The church at Atlantida was my faculty of architecture.*'

For those familiar with its distinctive curved walls the church has become an iconic representation of his work (*Fig. 4.2*). Like so many of his buildings the church appears at first sight to be rather improbable. How was it built? How does it stand? An exploration of form that avoids rational construction.

'*Architects..... like to be given churches as it gives them the opportunity to evolve strange shapes.*'[91]

On closer inspection it becomes clear that the church is part of Dieste's search for cosmic economy. A search which had now led him to the role of architect. The design of the church is a forceful statement by Dieste against what he describes as 'pseudo-rationalism.'

'*The rationalist who does not manage to produce architecture fails not because of an excess but rather because of a lack of true reason.*'[92]

Dieste here refers to the design of the building not solely in terms of its cost and construction, the aspects of its materiality, but more significantly in its symbolic, cultural and religious role, its metaphysical presence. In undertaking the design of the church, Dieste consciously straddled the commonly perceived

divide between engineers and architects - the debate between economy of means and architectural expression.

The church is situated in the village of Atlantida,[93] 40 km from Montevideo. The village is populated by manual and domestic workers, employed in the nearby spa town of the same name. The village itself consists largely of dispersed groups of houses and farms and lacks a clearly defined centre or focus. It is devoid of a sense of place: its position justified only by its proximity to the nearby more affluent town.

In 1952, Dieste was approached to act as structural engineer on a project to construct a

Fig. 4.1. Front elevation - Church at Atlantida.

Fig. 4.2. Front elevation.

simple church, commissioned by a wealthy benefactor, Alberto Giudice, who led religious instruction in the village. The background to the commission and Dieste's involvement are not well documented and Dieste himself prefers not to dwell on the details.[94] Juan Martin Piaggio however reports on six years of argument.[95] The essence of the argument between Dieste and the Giudice seems to have been on the nature of the building itself. Giudice wanted a simple, low cost building, 'The poor cannot afford beauty': whereas Dieste was insistent that the church should aspire to greater things, 'The poor deserve beauty'.[96] Dieste encouraged Giudice to appoint an architect and even provided a list of suitable architects; but to no avail.

'When we argued on the shape of the church, he argued that I should not worry so much about that matter due to the fact that our people were quite ignorant and had no aesthetic education.'[97]

A simple functional[98] building might have served adequately as a place for religious service but would have lacked any quality of aspiration for the congregation. Dieste's experiences on this project clearly had an influence on his own subsequent writings, where he frequently refers to the importance of self-fulfilment as a major part of human development. He believes that 'modest' people are sensitive and appreciative of art, architecture and beauty. Dieste also saw that the church could be a catalyst for the development and formation of a community space for the village. Giudice in making his generous donation,

Chapter Four
The reluctant architect

was still concerned about the cost and no doubt felt that an engineer without an architect would produce a simple and, hence, low cost building. He could look to the example of how engineers were alone appointed to construct purely 'functional' buildings. Dieste finally resolved the argument, largely by insisting that he be allowed to design the church and at the same time guaranteeing that the building would cost the same as a 'galpón'.[99] Work eventually started in 1958 and was completed in 1960. The church was built for $30 per square metre (approximately £20 / m²). The project became an obsession for Dieste. He received no fee for the design and worked on site during the day, leaving in the evening to earn his living working for Viermond.

The beginning of the twentieth century saw the growth in a grassroots movement for change in the Catholic Church. The movement promoted spiritual renewal and re-evaluation of the role of the laity[100] in the activities of the church. The traditional Tridentine mass was seen as too ritualistic, as a ceremony where the priest performed in a mysterious way, using a language that most did not understand and with the congregation as passive observers. Pressure grew for the mass to return to its origins as a congregational celebration of Christian worship. By the middle of the century the movement, now formally recognised as the Liturgical Movement, had spread across many countries, and the hierarchy of the Church responded to this call for change. At its heart the Liturgical Movement strove for:

'...making the liturgy known loved and practised in a better fashion, particularly by the full participation of Christians in the liturgical celebration.'[101]

Pope Pius XII (1939–58) encouraged and endorsed the movement, and it became an important activity of the Church proper. The reforms were finally enshrined in the 'The Constitution of the Sacred Liturgy' issued by the Second Vatican Council, presided by Pope John XXIII in 1963. The liturgy became a celebration of the Christian faith with much greater participation by the laity. The mass would now be performed in the vernacular language of the country.[102]

The Liturgical Movement had a great influence on church design. After the Second World War there was a tremendous programme of church building, to replace the many that had been destroyed. For the Church this was an opportunity to deal with the ideas of the Liturgical Movement and to show that religion was evolving in tune with changes in society. Dioceses were open to, and even encouraged, radical designs that challenged traditional ideas on church plans and used modern materials, in harmony with changing notions of the function of the building.

'The result (of the Liturgical Movement) was that the 1950s and early 1960s were an exceptionally inventive time for church architecture.'[103]

Although Dieste was aware of the debates of the Liturgical Movement these endorsed rather than sculpted his ideas. He was clearly concerned with the needs of the Church itself; but, more fundamentally, concerned to ensure that the Church had at its heart a true sense of being for the congregation. His understanding of the Church in the community was so strong that he would always have striven to design the church in his own way, as an expression of his sensitivity to the human spirit and perhaps a reaction to Giudice's well-meaning but insensitive benevolence.

Atlantida was his first architectural commission and, having no formal training in architecture, he struggled initially to develop a design that fulfilled his aspirations for what the church should be. Dieste attacked the problem with the same determination that he used in developing the complex mathematics of Gaussian vaults. The need to bring together the various strands of building performance and architecture, so often at odds with each other in so many contemporary buildings was critical to satisfy both the benefactor and his own ambitions. The architectural critic Marina Waisman,[104] has described the work of Dieste as being open to interpretation in different ways; by the economist, the sociologist, the technician and the spectator, in terms of cost, function, efficiency and beauty.[105] Atlantida can be read by each of these 'observers', with singular clarity. The church is constructed as a single volume delineated by the repetitive cross-section of undulating wall and roof. At ground level the walls follow a straight line, enclosing a rectangular plan. From here the walls, undulate as they rise, describing a parabolic curve in a horizontal plane. The amplitude of the curves increases to a maximum at the eaves. The walls incline both inwards and outwards creating a spectacular internal effect and a unique internal space. The generous undulations of the walls and roof wrapping and folding around the congregation like a rich, thick blanket, accentuated by strong but soft edged shadows provide a sense of security and warmth within the building.

The building is relatively small; 16 m wide by 30 m long and 7 m to the eaves. It is perhaps a mistake in the design of many small churches that they mimic the plan of larger churches, incorporating all the features of larger churches including baptismal font and separate side altars and aisles. Often these 'mini-cathedrals' simply do not work, either visually or functionally and instead draw attention to their smallness.[106] In Atlantida the impression is quite different, the church feels larger than it actually is. The inclination of the walls heightens the perspective from the entrance to the altar. At eye level the building seems greater than its footprint.[107] There is an absence of the conventional structural elements of churches - columns, groins and arches - which are used to divide the internal space. Instead, they are replaced entirely by an undulating organic structural form that encloses the entire building. The single volume open space allowed Dieste to create an integrated plan which deliberately brought the priest closer to the laity.

Fig. 4.3. Altar and nave.

The nave and the altar share the common space. Around the altar is a free-standing U-shaped wall that opens towards the nave, connecting the two spaces (Fig. 4.3). In fact, the internal plan of the church is such that on entering the only focus of attention is the altar and the nave, the two primary spaces of the church used during the mass. The ancillary activities such as the confessional, sacristy[108] and baptismal font are not apparent on entry. The curved wall around the altar has an important programmatic function in breaking the traditional hierarchy between the priest and the laity. In a conventionally planned church, at the start of the

Chapter Four
The reluctant architect

mass, the priest would enter the altar directly from the sacristy.

'In all of the churches that I knew, the priest appeared suddenly, a little like the doll in a jack in the box.'[109]

The altar chancel is divided from the nave by a small wall or rail. The priest enters the altar and would remain there throughout the service, separate and distant from the congregation. In Atlantida, the entrance of the priest is now processional. The sacristy is placed behind the altar, physically separated from it by the wall. At the start of the mass, the priest walks around the side of the altar addressing the congregation directly, before entering the chancel from the

Fig. 4.4. Altar for Chapel to Our Lady.

Fig. 4.5. Baptistry and entrance.

nave. Thus, the priest approaches and perceives the altar from the same perspective as the congregation. At Communion the congregation approach the steps of the altar, between the curved walls. To Dieste, this is symbolic of a return to the origins of the mass as a celebration of the Last Supper.

At the side of the chancel, tucked behind the curved wall, is a small chapel to Our Lady of Lourdes. The altar for this chapel is a prism of brick, projecting through the side wall (Fig. 4.4). The prism tapers from inside to outside and the bricks themselves are cut into tapers to accentuate the perspective. Diffuse light passes through an onyx window at the end of the prism. The chapel is neither visible nor signalled from the nave, creating a very quiet, spiritual space for contemplation and meditation.

In the Catholic Faith, baptism represents entry into the spirit of the Church. It is imbued with a sense of ritual concerning rebirth. Traditionally, at the beginning of the Christian era, its roots lay in the conversion of non-Christians, where converts were cleansed of original sin and 're-born'. In the contemporary church, most baptisms are performed on the newly born with the parents and god parents participating, re-affirming their faith and attesting their commitment to the spiritual development of the child. Dieste felt that much of this ritual had lost its significance, and took the opportunity to bring added meaning in a poetic and spiritual way. The baptismal font is situated outside the church in a circular underground chamber, directly in the centre of the circle and is lit from above through a circular window of translucent onyx, the only feature of the baptistry visible from the outside. The entrance to the baptistry is a triangular prism of brick piercing out from the ground in front of the church, concealing a staircase down to the baptistry (Fig. 4.5). The ceremony commences with the priest welcoming the family and child at these stairs and then accompanying them down to the baptistry. After the ceremony, the priest and family enter the church through a tunnel and staircase upwards directly into the nave, an evocative passage representing resurrection and re-birth into the church.[110] Dieste's ideas on the baptistry were actually quite different from the arrangements

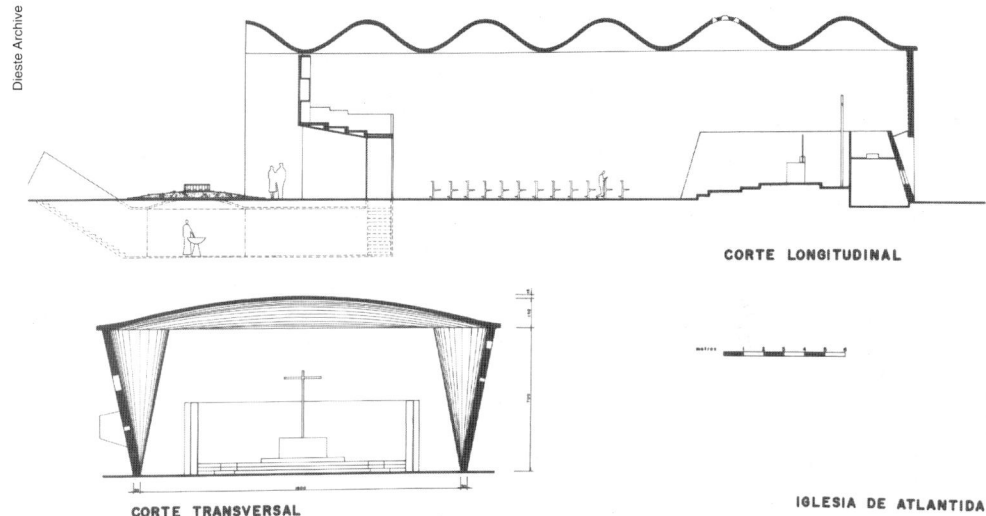

Fig. 4.6. Sections.

Chapter Four
The reluctant architect

Fig. 4.7. Rear of church.

Fig. 4.8. Stairs to choir.

Fig. 4.9. Construction of walls.

suggested by the second Vatican Council that suggested that the baptismal font should be placed much closer to the centre of the congregation to allow the full congregation to take part in the sacrament.[111]

On the front elevation the curved side walls incline outwards and the roof curves upwards, taking a cut across the undulating section at its widest point (Fig. 4.6). The façade, on this elevation sits inside the section, set back from the edge by one half wave of the curve, and so positioned at the narrowest point of the section. The section thus expressed on the front elevation appears as if it has been sliced through and plugged with the front façade. The effect is of a large portico that opens to welcome the congregation. The front elevation itself is split horizontally into two distinct parts. The lower part, at ground level, contains the entrance to the church, set back underneath the choir loft on the upper level (Figs 4.7 and 4.8). The upper part of the façade is constructed using three rows of inclined prefabricated brick panels forming a series of staggered vertical louvres. The gap between each panel is glazed with translucent sheets of onyx. Light is deflected and diffused through these louvres. An intense pale glow is created inside. The confessionals are tucked underneath the choir.

Conceptually, the structural form is straightforward but nevertheless sophisticated in execution. The side walls and the roof can be read as a two pinned portal frame that maintains its lateral stability by the rigidity of the connection

Fig. 4.10. Walls constructed to eaves.

Fig. 4.11. Form work for roof.

Fig. 4.12. Form work in place.

between the wall and roof. The section is repeated four and a half times along the length of the building. The lateral pressure caused by wind is the primary loading condition on the wall. The form of the wall is a reversal of contemporary ideas about tall, single storey masonry walls. In the UK such walls tend to be designed as free cantilevers with rigid connection at the foundation, transferring the lateral bending moments to the foundation. The roof to wall connection is considered, at best, as providing a prop to the top of the wall. Curtin et al. in their book, *The structural masonry detailers manual*[112] describe various forms of improved geometry walls suitable for tall single storey buildings as being more efficient in resisting lateral loads.[113]

'*The bulk of masonry walls are built by tradition as plain walls (and therefore act structurally as plate sections) but the interest shown in the authors' development of diaphragm and fins has heightened the approach of other geometric and more structurally efficient sections.*'

Atlantida pre-dates this statement by more than two decades and in terms of structural efficiency is a further step ahead. The sections developed by Curtin are constant throughout their height even though the bending forces diminish towards the top of the wall, whereas the walls in Atlantida taper in section according to the bending moments. A further advantage of the form is the simplification of the foundations which are sized according to the thickness of the wall. In a wall of constant cross-section the foundations are, by necessity, much wider, based on the overall width of the wall. Possibly up to six times wider than necessary. The foundation may be constructed to follow the profile of the wall requiring more complex foundations.

The walls of Atlantida are constructed in 300 mm thick brickwork, reinforced with 3 mm diameter steel wire placed in a cavity between the inner and outer leafs (Fig. 4.9). The curved form of the walls, may appear more suited to the fluidity of concrete construction, particularly in the Latin America style of Candela and Niemeyer. However, the form is actually achieved more easily using brick (Fig. 4.10). Dieste fully exploits brick and avoids the complex formwork needed for a concrete wall (Figs 4.11 and 4.12). The wall is constructed on a reinforced concrete ground beam, supported on bored piles at 3 m centres.

Chapter Four
The reluctant architect

The wall is set out by two sets of lines. The plan (*Fig. 4.13*) shows both of these lines; a straight base line at ground level and parabola defined in a horizontal plane at the underside of the roof. The surface of the wall is described by drawing a series of straight lines between the baseline and the roof line. Moving along the baseline each vertical line is therefore displaced to meet with the parabola at the eaves. The sequence of straight lines define the undulating surface of the walls. During construction, the bricklayers follow the pattern of string lines stretched between the base of the wall and a timber scaffold at the eaves. The bricks are laid in a conventional manner following an adapted form of stretcher bond. The bed joints follow the inclination of the wall surface, which is continuously changing, and they therefore rotate along the length of the wall. Laying the bricks in this manner requires considerable skill to continually adjust the mortar bed to align the bricks correctly, but it avoids stepping and corbelling the bricks. The bed joints play an important role in the visual expression of the wall, maintaining continuity and the sense of a folded surface. If concrete had been used then the construction process would have been more complicated due to the formwork but, just as important, the expression of the wall would also have been entirely different. Economics normally demand repetitive uses of formwork. Between each cast there is the inevitable construction joint, impossible to hide convincingly, changing considerably the appearance of continuity and undulation obtained. The bed joints provide continuity of the undulations in the brickwork. In Dieste's opinion walls are best constructed in brick.

' *Concrete is ideal when it's made into a slab. If it is used for vertical elements it becomes more expensive.*'[114]

This view is supported by the use of 'tilt up' concrete construction techniques,[115] developed in the USA to simplify the formwork and construction of vertical planar concrete elements.

The roof is a double-curvature Gaussian vault connected to an edge beam at the top of the wall, The roof spans between the side walls between 13·2 and 18·8 m as they follow the

Fig. 4.13. Plan.

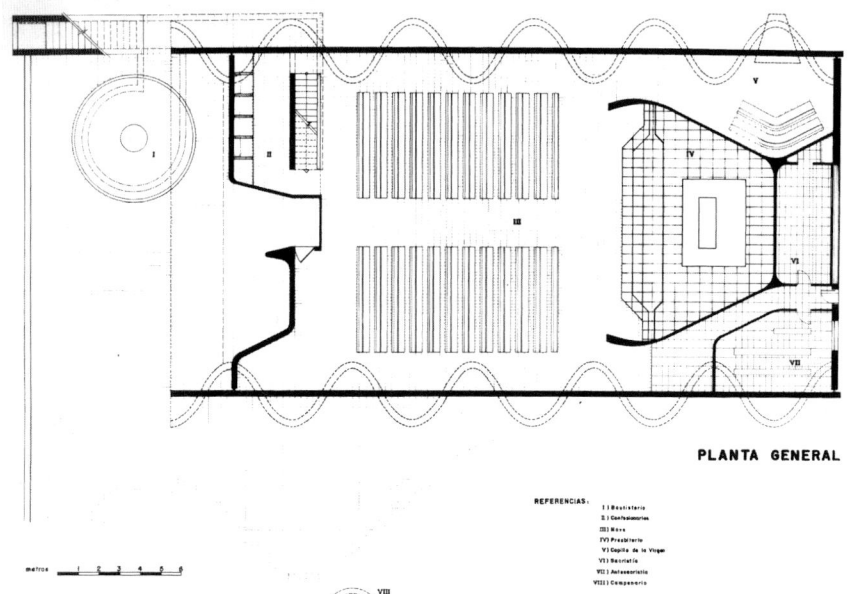

IGLESIA DE ATLANTIDA

Fig. 4.15. Edge beam.

undulations at the top of the wall. The rise of the vault varies as the span between the walls changes; from a minimum of 70 mm at the shortest span to a maximum 1·47 m at the widest section. The structural design is based largely on Dieste's intuitive appreciation of the physical behaviour of the roof. He had not yet formalised his thinking in the mathematical theory. The vault can be thought of as consisting of two separate segments, one acting as an arch and the other as a tie. The segments where the rise is higher form shallow arches between the side walls which, if adequately braced, will generate compression in the roof. The shallower segment of the vault, at the shortest span the vault is almost flat, and is not capable of developing arch behaviour. This segment is 'suspended' from the arch segments with steel reinforcement. The lower segment contains a steel tendon that restrains the horizontal thrust of the roof. In other Gaussian vaults, such as the warehouse at Montevideo Docks,[116] the rise of the lower segment is much greater. However the ties are exposed; an unacceptable condition for the church. Each tendon consists of four large steel bars, typically 32 mm in diameter, which are anchored into a 12 cm thick reinforced concrete edge beam sitting on top of the wall. The edge beam is a primary element in the construction and resolves a number of problems.

- It transfers the thrust from the vault to the tendon.
- It stiffens the top of the wall.

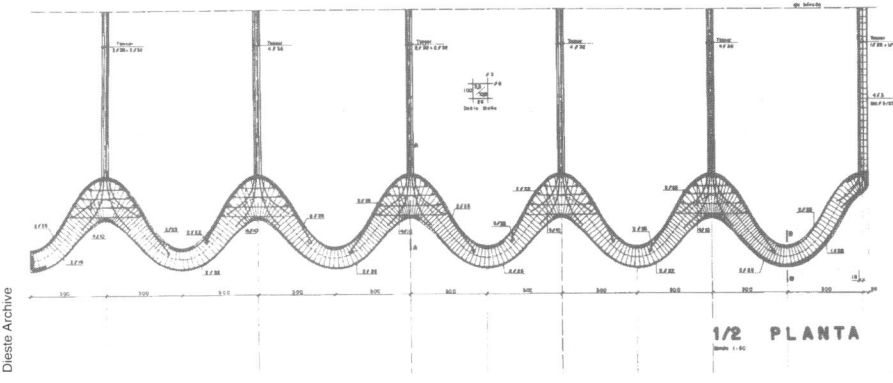

Fig. 4.14. Part plan section at eaves showing tendons.

- It provides a horizontal surface that reconciles the intersecting geometries of the wall and roof. The edge beam traces the curvature of the wall and cantilevers outwards. It varies in width from 1·7 m at the anchorage point to 0·83 m, one half wave along the wall where the span is at maximum. The width of the edge beam varies according to the bending induced by the thrust from the roof. This can be seen clearly in the simplified diagram of the edge beam and tendon (Fig. 4.14). The soffit of the edge beam is finished with 3 cm thick bricks. In the design of the edge beam Dieste again creates a detail which is complex but is rational in the way that it resolves a series of technical and architectural problems. It could have been simplified considerably by adopting, for example, a rectangular plan spanning across the top of the wall, easier to analyse, design and construct. If this apparently rational solution had been accepted[117] the form of the roof would have been compromised. The link between the crests and troughs of the roof that coincide with the peaks and troughs of the wall would have been broken, the roof terminating in a flat soffit. The edge beam is a hidden detail, not visible from inside the building and appearing as a continuation of the roof externally (Fig. 4.15). Its form can only really be appreciated from the diagram, which reveals an organic character of bone and sinew. Even the steel bars are curved, rooting themselves firmly into the concrete of the edge beam.

The side walls and roof were completed before the front and back walls (Fig. 4.16). The end walls are free standing, independent of the side wall and roof, closing the open ends of the church. The walls are not connected to the side walls, a gap runs between, again glazed with translucent onyx. The decision to make this separation has both a practical and architectural logic. Practical, in that the bonding of the two walls using brickwork is difficult to achieve

Chapter Four
The reluctant architect

Fig. 4.17. Windows on front elevation.

Fig. 4.16. Side walls and roof.

satisfactorily for the bed joints of the inclined walls would always be out of alignment with the horizontal joints of the front walls. Architectural, by expressing the curvature and inclination of the side walls on the external elevation and creating dramatic internal lighting. Internally, Dieste shows an effortless understanding of light, one that helps create the spirituality and mystery of the church. The side walls are punctured with small coloured glass windows at high level (Fig. 4.17). The windows range in size in modules of between two and six bricks and the glass, which is glazed directly into the brickwork, is designed to have minimal interruption on the surface of the wall.

The rear wall creates the backdrop to the simple wooden crucifix behind the altar. The bricks in the rear wall are laid with the long edge canted, outwards and the bonding pattern is staggered to produce a deeply textured surface that reflects light from underneath (Fig. 4.18). The edges of the bricks appear as little tongues of flame. Here Dieste alludes to an important symbol in the Catholic Church, representing both the soul, with warmth and radiance, and also the Risen Christ as the light that sprang from the darkness,[118] a direct consequence of the Crucifixion. Light is reflected upwards from an inclined wall below, roughly the size of a football goal, cut out and folded from the rear wall.

Above the altar, small circular windows are arranged around a larger roof light (Fig. 4.19). The windows are formed of ceramic tubes[119] built into the roof, precisely aligned with mortar joints. The central roof light fits exactly into a square described by eight bricks. The surrounding lights are centred on the intersection of perpendicular joints.

The roof itself is a beguiling exploration of geometry and a product of Dieste, the engineer and constructor. The consistency and accuracy in the positioning of the bricks (which can also be seen in the church of San Pedro in Durazno) has generated a series of intersecting lines that illustrate both the form and conception of the roof. The lines running between the side walls follow the catenary directrices of the roof. The joints are a graphical expression of the roof geometry, deliberately executed; they appear like the transformation of a three-dimensional wire frame model into solid surface. The accuracy of construction is also evident in the roof to wall

Fig. 4.18. Rear wall with canted bricks.

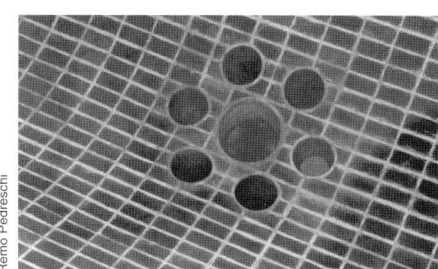

Fig. 4.19. Window over altar.

Fig. 4.20. Junction between roof and wall.

junction (*Fig. 4.20*). There are no transitional details to cover where the wall and roof meet (*Fig. 4.21*). The conventional solution (conventional because it is simple) demands an interface zone, either an edge condition that can be realigned and levelled to deal with the accumulation of tolerance or some form of capping piece, such as a cornice, to hide any irregularities at the interface between wall and roof. The brick bonding pattern of both the roof and the walls does not alter at the intersection, an apparently simple detail. However, reflecting on the process of construction a greater significance can be discerned. The walls are constructed first, up to the underside of the roof. The roof is then assembled on a prefabricated formwork supported off the side walls. Thus, the top of the

77

Chapter Four
The reluctant architect

Fig. 4.21. Roof and wall.

Fig. 4.22. View from Chapel to Our lady.

wall sets out the line and level for the roof and dictates the quality of the junction. Here, Dieste is pushing the construction technique to the limit, but with confidence in both his own mathematical skills and in the skills of his workforce. The undulating wall has to be accurate in finished level to no more than one bed joint, say 10-15 mm over a length of 30 m (Fig. 4.22). This is not the product of an anally retentive technocrat but the reflection of an artist trying to achieve that which is both possible and worthwhile.

Returning to Waismann's analysis, the church has shown all the characteristics she attributed to Dieste's work generally. At least two architectural commentators have drawn comparison with Le Corbusier's church at

Ronchamp. Juan Martin Piaggio[120] compares the plan of the church to the spiral of the Modulor and highlights similarities between the front elevations. Karl Ludwig Diehl considers similarities in the lighting but suggests that Atlantida is a much clearer construction, being completely exposed compared with the rendered masonry used for the walls in Ronchamp. The arrangement of windows in the side walls has a Corbusian feel. These may be valid comparisons. They do not, however, prove Dieste's indebtedness to Le Corbusier at Ronchamp. Dieste was not aware of Ronchamp at the time when he was designing Atlantida.[121] Rather than similarities in particular details perhaps the most significant comparison lies in a shared understanding of the significance of natural light.

'The key is light and light illuminates the shapes and the shapes have an emotional power'. Le Corbusier[122]

The construction of Atlantida shows a delight in its making, developed to express the form of the structure using surface, texture and light. In one sense, its construction and design is also a metaphor for his attitude towards what the Church should be, a Church for the people that takes pleasure in the familiarity of a local material, an every day material, a modest material constructed using a predominantly local labour force,[123] to create an inspiring form, almost not of this world.

'These materials, however, were handled with a care that aspires to create the homage that these humble people deserve.'[124]

The people of Atlantida have grown to love this building. Dieste tells of the great pride he felt on overhearing an old peasant women proudly showing her friend the church and explaining with great detail and insight the intentions in Dieste's brief.

Chapter Five
In the footsteps of tradition

Chapter Five
In the footsteps of tradition

Fig. 5.1. Original façade.

Fig. 5.2. Durazno.

In 1967, a fire in a parish church provided Dieste with the opportunity to complete his second church (Fig. 5.1).[125] Although built some ten years after Atlantida, the church of San Pedro carries forward some of the ideas Dieste had been pondering at that time. The need to follow contemporary thinking about the role of the Church in the liturgy, bringing the congregation and clergy closer together,[126] opening the space and setting the altar closer to the laity have been important in shaping San Pedro. There are also similarities in the conceptual process, with Dieste again extending and developing the original brief and preparing a much bolder and more challenging technical and architectural proposition than originally intended. However, its design and construction are different, responding to a distinct set of conditions. The church is situated on a tightly bounded site, surrounded by other buildings. Consequently, the reconstruction followed the original footprint. Nevertheless, Dieste produced a design that exemplifies his holistic attitude to form and construction. Compared with the spectacularly curving walls and roof of Atlantida the flat planes of San Pedro may appear to be more reticent and restrained. The structure, nevertheless, responds to Dieste's principle of the resistance of force through the use of form to produce a carefully designed folded plate solution of apparent simplicity. In many ways the restrictions of the brief have helped to produce a potent and spectacular internal space.

Durazno is a small town, the capital of the region which bears the same name, about 180 km from Montevideo, heading towards the centre of Uruguay through sparsely populated, flat countryside. The total population of the region is less than 60 000. The area derives its income from agriculture, cattle, sheep, wine production and wheat. The town itself is laid out in a repetitive rectangular grid of fairly wide, tree lined streets bordered by mostly single storey, flat roofed buildings (Fig. 5.2). Responding to this rather odd scale, many of the trees are cropped to the height of the buildings. The church and an incongruous apartment block[127] next door are the tallest buildings in the town. San Pedro is situated on the south side of Plaza Independencia, the principal public space in the town. The casual visitor to Durazno is unlikely to come across the modern church hidden here. The only remnant of the original church and its only public façade is the Colonial entrance and bell tower, facing onto the square. Tucked away in this small sleepy town, far from Montevideo, it is not surprising that the building is less well known than Atlantida.

The original timber roof was completely destroyed in the fire. Dieste, together with the architect, A. Castro, and the engineer, R. Romero, were consulted on its repair. Two possible solutions were identified.[128] One was to repair the existing wall and replace the roof while maintaining the original character of the church. The alternative was to produce a new roof of distinctive character and avoid using the original walls of the nave as load-bearing. Dieste favoured the latter approach for various reasons. The original stucco had never been of particularly good quality and repair was not justified; and some of the columns that lined the nave had cracked during the fire and were of dubious strength. Perhaps more importantly, there was the opportunity to adapt the church to the demands of the new liturgy.

Dieste's view prevailed, thus the only part of the church that was retained was the entrance façade and the atrium. Given that he had such success with Atlantida and at such low cost, it is worthy of comment that he should retain this relatively small part rather than argue for complete rebuilding. Dieste[129] restricted his proposal for two reasons. Firstly, the existing façade was the principal focus of the major public square in Durazno and he wanted to maintain its historical

Fig. 5.3. Altar.

Chapter Five
In the footsteps of tradition

Fig. 5.4. Handmade bricks.

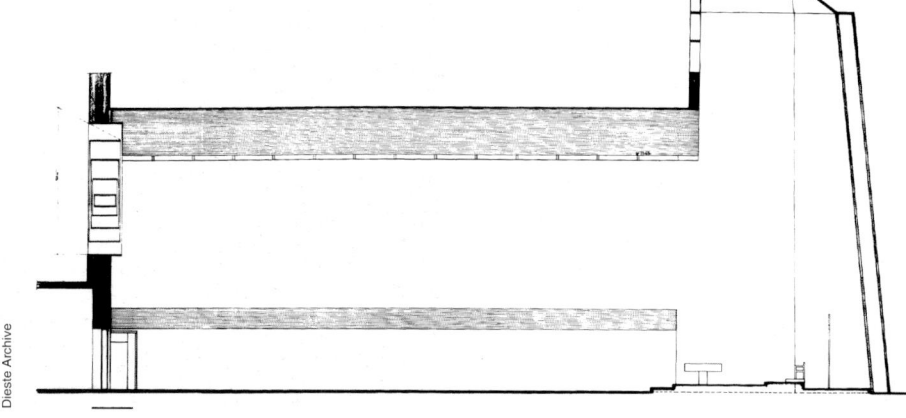

Fig. 5.5. Long section (without entrance).

importance. In Atlantida, he had felt that there had been a missed opportunity for the church to play a part in the development of a new civic public space. Secondly, he felt that he had to show respect for the workers who, only a few years previously, had made great efforts in an earlier renovation of the church. Many of whom were still alive and living in Durazno. Dieste was sensitive to the preservation of both human and cultural values.

The sides and back of the church, closely surrounded by other buildings, mostly small houses, gave little opportunity for external expression or expansion. The decision to retain the façade effectively directed the organisation of the plan and set up the primary axis of the church. The reconstruction, therefore, closely followed the original footprint. The walls of the nave and presbytery were demolished; but the lower walls to the side aisle were retained and lined internally with a new skin of brickwork, canted inwards from its base. New nave walls, a large tower behind the presbytery and the roof were added. What were once the side aisles are no longer separated by colonnades but form one continuous space with the nave.

There are many structural innovations in this building which are not apparent on first entering. The visitor is immediately drawn to the dramatic use of controlled light as it reflects on the side walls of the tower, illuminating the suspended crucifix, which, tilted at a slight angle to the vertical appears to be rising upward towards the source of light, in almost literal representation of the ascension (Fig. 5.3).[130] There are no windows puncturing the nave walls, which are inclined inwards to provide a heightened sense of perspective. Light is drawn in through continuous strip windows between the top of the walls and the roof. The roof appears to hover just above the walls. The strips of strong light together with the inclination of nave wall [131] add further emphasis and focus to the altar and the tower. Handmade bricks in a rich reddish brown colour are used throughout to provide continuity to the internal surface (Fig. 5.4.). The walls and roof, folding round the internal space, seem to have been formed by some clever origami. The bricks are laid in a stack bond pattern with continuous vertical joints, subtly suggesting that the building may not have been constructed by purely traditional methods.

Indeed, there is much more than meets the eye. After a short while, the visitor notices that there are no columns supporting the 8 m high nave wall, now spanning a full 32 m from the front to the back of the church (Fig 5.5). Nor are there any columns supporting the brick roof on its point of departure from the nave. There is no formal expression of the structure in any commonly understood architectural language associated with planar structures of this scale. Where is the structural frame or beam? To span such a large opening one might reasonably expect to see a large beam, sitting on a correspondingly large padstone, in turn supported on a substantial column. But not here. The brick must simply be a skin to a hidden structural frame! In traditional churches and even many contemporary churches of the Liturgical Movement, the expression of the ribs of structure or frame create implied boundaries between the spaces. In San Pedro the brickwork is structural and expressed both in the roof and the walls and the bonding is continuous between these elements, such that the roof becomes the wall and the wall becomes the tower. The dark side aisles, the colour of the brickwork, the light entering from high level, the inclined walls of the nave providing the gradation in light, all create the impression of a vast cavern burrowed in the fertile Uruguayan soil (Fig. 5.6).

In reality the walls and roof are rather thin, as revealed in the section (Fig. 5.7). There is nothing precocious or frivolous in this approach. The avoidance of formal structural expression and the continuity of the surfaces are essential elements of the architectural programme to create a single, unified space. Dieste would probably argue that he does have an expressed structure. In his terms, he is creating structure through form and its formal expression is the folded plate. The folded plate is a structure associated typically with steel or concrete, each of which has a quality of plasticity or fluidity, steel by virtue of its ductility[132] and concrete by its formlessness during casting before its final monolithic state. Plasticity or fluidity are not characteristics easily attributed to masonry. Masonry construction is both sequential and incremental providing a visual history of its making. For example, the keystone is obviously the last stone to be placed in an arch.

Fig. 5.6. View from side aisle to altar.

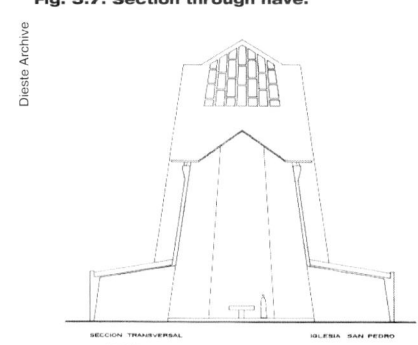

Fig. 5.7. Section through nave.

Chapter Five
In the footsteps of tradition

Fig. 5.8. Junction between nave wall and roof.

Fig. 5.9. Junction between nave, presbytery and side aisle.

Many of Dieste's buildings appear plastic, as if deformed from an initially flat surface.[133] Apparent plasticity is obtained by exploiting the incremental sequence of brick construction, often by using a form defined by a series of undulating straight lines to generate a continuously curving surface. The success of this approach is clear in Atlantida. In San Pedro the same sense of plasticity is there, but it is less obvious and in some ways more complicated. The walls of Atlantida, although brick, suggest the use of formless concrete, while the flat planes of San Pedro could easily be perceived as a folded steel plate structure. In a steel structure plasticity comes from the microstructure of the material itself, allowing it to be bent without fracture and is conspicuous only at the junctions between planes.

The primary difference between the two churches and their use of brickwork lies in the way the surfaces change. At Atlantida change is continuous, whereas at San Pedro change is abrupt.

In traditional masonry construction, junctions between surfaces are most easily dealt with using either simple angles or creating special shapes of bricks or stones. In Dieste's work, the use of special bricks or lintels to smooth the junctions between planes would differentiate and emphasise discontinuity, therefore denying both the expression of the structure through its form and the continuity of the space through the internal surface of the building. A remarkable aspect of the San Pedro church is the joining of the surfaces. Dieste does not use special bricks; the bonding pattern and joints between bricks run continuously between planes in an almost seamless fashion, as if tailoring a fabric to a form. The bricks are laid in a stack bond arrangement that helps to reconcile the complex geometry of the surfaces that meet at particular points. The junction between the walls of the central nave, the presbytery wall and the roof and walls of the side aisle is a profound demonstration of seamless construction, particularly bearing in mind that the nave walls are actually heavy long span structures (Figs 5.8).

'The coherence between what the form shows us and the constructed reality is also important. Coherence makes the form intelligible to us.'[134]

The four inclined planes meet at a junction without a transitional node point, a padstone, special shaped bricks or even a cut brick (Fig 5.9). The successful execution of this detail requires care, devotion and exceptional precision in construction. The bricks in the wall at the head of the side aisle are laid at the same angle as the inclination of the nave walls to accommodate the bonding patterns of the nave. The vertical joints in the nave walls align smoothly with the vertical joints in the walls of the presbytery. Precisely because there is no apparent discontinuity in the surfaces at this junction, which would otherwise draw one" eye, a second look is required to truly appreciate its significance, particularly when one considers the complexity and magnitude of forces being transferred at this point. Vittorio Vergalito, Dieste's foreman in charge of the construction, on being complimented on the quality of this particular detail, looked bemused as if these things are self evident and replied[135]

'What else can I do? If I make a mistake it is a mistake for 100 years.'

The structure that Dieste designed evolves entirely from the architectural programme using brickwork structurally throughout. Many of the elements themselves were innovative and well ahead of international developments in structural masonry at that time.[136] The reconstruction of the building comprises the addition of the following elements: the roof over the nave, the nave walls and roof over the side aisles, the new presbytery. A concrete portal frame at the entrance to the presbytery is used to support the new roof and walls (*Fig. 5.10*). The roof and walls over the nave are both reinforced and prestressed,[137] using the overlap loop method, similar to the technique

Fig. 5.10. Concrete portal to presbytery.

used to prestress the valleys between the free-standing barrels vaults.

The roof over the central nave is a *tour de force* in the use of form to resist force. The inclined part lies at 34° to the horizontal, spans 7·38 m across the nave and is only 80 mm thick, comprising 55 mm thick bricks and 25 mm of concrete topping. Reinforcement is installed in the bed joints to resist bending. The section relies on horizontal reactions at each end to maintain geometric stability. An obvious problem with inclined roofs is how to provide adequate horizontal reaction to prevent the roof flattening out. The simplest solution is to tie the ends of the roof together preventing lateral movement. Dieste discounts this option, probably for two reasons. The first objection is primarily architectural. The

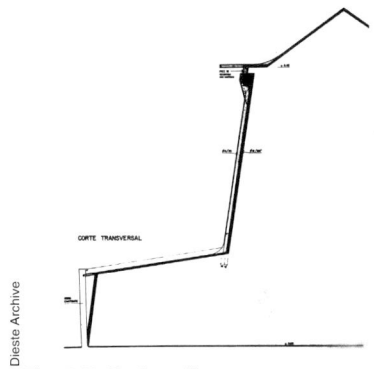

Fig. 5.11. Part section.

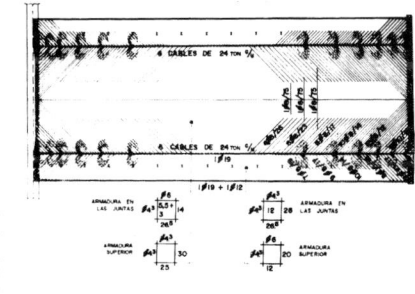

Fig. 5.12. Prestressing in roof.

Chapter Five
In the footsteps of tradition

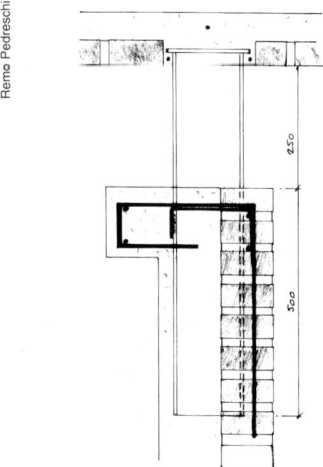

Fig. 5.13. Junction between roof and wall.

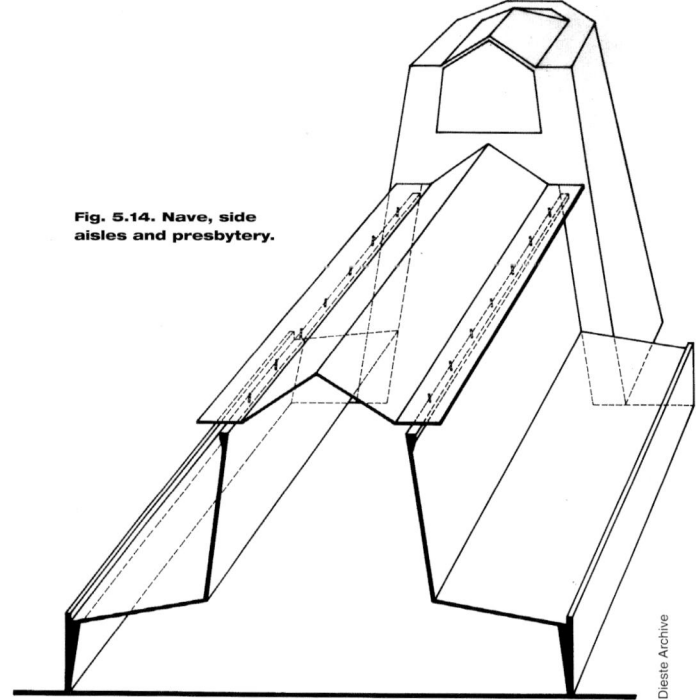

Fig. 5.14. Nave, side aisles and presbytery.

IGLESIA SAN PEDRO (Durazno)

tie would again act against the sense of uniformity of space, suggesting a dividing plane between roof and wall. The second reason is an exercise in the purity of his structural philosophy: the resistance of force through form.[138] He uses the form of the roof itself to resist the horizontal thrust.

The horizontal edges of the roof are 1·8 m wide and 120 mm thick, creating deep horizontal beams which also absorb the wind force on the walls and transmit it back to the end wall and portico. The horizontal slab overhangs the nave wall (*Fig. 5.11*), creating a shading device that prevents the leakage of direct sunlight into the church, sharpening the light through the horizontal slits between roof and wall. The depth of the roof itself, from eave to ridge, spans the full 30 m of the nave. The roof is prestressed using staggered looped tendons (*Fig. 5.12*). The tendons consist of six groups of 16 mm diameter bars. The arrangement of the tendons follows the bending moment, progressively curtailing each tendon from the middle of the span to each end. The total prestress force is 142 tonnes and combines prestressing to counteract both the horizontal thrust in the roof and the vertical bending in the roof form.

In calculating the prestress, Dieste considers the forces acting in each of the planes of the roof elements and then determines the level of pre-compression necessary to counteract the tensile stresses. The total prestress is the summation of the force needed in each plane. The prestressing cables are placed at the junction between the horizontal and inclined planes of the wall. Here the thickness of the concrete topping over the bricks is increased to accommodate the prestressing cables. The roof also supports the nave wall (*Fig. 5.13*) against lateral loading and is connected using steel stub beams built into the top of the wall forming the gap for the strip window.

The inclined nave walls also span 32 m from the concrete portal onto pillars at the entrance to the nave (*Fig.5.14*). The wall is constructed as a composite concrete and brick wall, 270 mm thick. The wall is thickened locally at the top to incorporate a steel stub column which props the horizontal plate of the roof and transfers the wind load from the nave to the roof. In addition to its own weight, the wall also supports the roof over the side aisles. The total

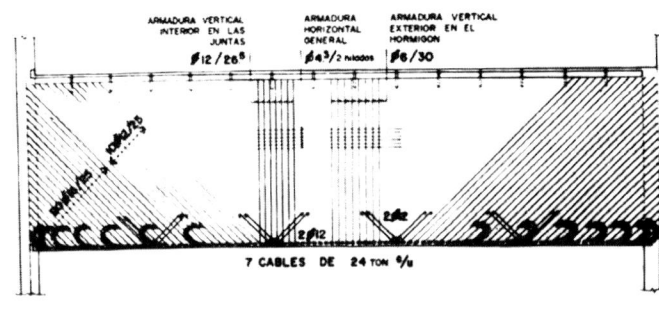

Fig. 5.15. Prestressing in nave wall.

Fig. 5.16. Presbytery from rear.

Fig. 5.17. Plan section through presbytery wall.

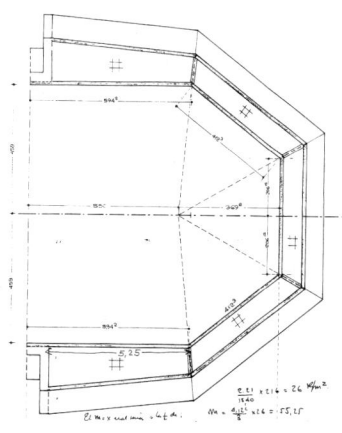

load is approximately 200 tonnes. This large span and large load generates very high bending moments and again prestressing is used with the same techniques as in the roof (*Fig. 5.15*). Seven tendons producing a maximum prestress force of 169 tonnes were installed and curtailed to follow the bending moment diagram, reducing in stages towards the supports. The prestressing is contained within a thickened section of wall, at the junction between the nave wall and the aisle roof. The aisle roof shares this pre-compression and provides lateral resistance to the wind pressures on the nave. Thus the prestress in both the roof and the walls is applied at the junction between two planes, each plane contributing stiffness and resistance where most appropriate to carry either vertical or horizontal loads. The loads from the nave roof and wall are transferred to a reinforced concrete portal frame at the entrance to the presbytery and to brick-clad concrete columns at the entrance to the nave. Diagonal steel bars carry the large shear forces in the wall back to the concrete portal.

The roofs over the two side aisles are of ribbed concrete and reinforced brick construction. A series of concrete beams, at 5.8 m centres span from the original side walls to the nave. A reinforced brick slab, 120 mm thick spans in two directions between the beams and the walls. The soffit of the slab is completely brick, with concrete beams projecting outwards.

The other major element of the reconstruction, the presbytery tower, is constructed as a reinforced brick diaphragm wall

Chapter Five
In the footsteps of tradition

18·4 metres high and inclined inwards from its base (*Fig. 5.16*). The wall consists of two skins of brickwork, varying between 1·0 and 1·6 m apart, connected by horizontal and vertical ribs of brickwork (*Fig. 5.17*). Each skin of brick is reinforced to resist wind pressures between the ribs.

In the use of structural masonry the church has many innovations and it is worth pausing to consider the situation in masonry research and applications around the time. In the UK, from the late 1960s through to the 1980s, there was great activity and interest in the use and potential of prestressed and reinforced masonry, probably more than in any other part of the developed world. Research was underway in a number of universities and well known structural engineering practices like, W G Curtin and Partners, Bradshaw, Buckton and Tonge and Harris and Sutherland among others, were designing masonry structures using prestressing and reinforcement. In a thorough paper to the Institution of Civil Engineers,[139] James Sutherland described the current state of structural masonry in the UK. While his comments on un-reinforced masonry and its application in high rise buildings showed a strong sense of conviction for the material, he was rather more circumspect on the potential for reinforced and prestressed brickwork. He suggested that reinforced masonry would compete with concrete in only a few particular areas, namely retaining walls and similar structures.

'*One reason frequently given for the scarcity of examples of reinforced masonry in the UK is the lack of an adequate code of practice. If this is so it is a sad reflection on our independence of action as engineers.*'[140]

Scarcely a criticism that could be levelled at Dieste. Dieste has never seen reinforced brickwork as a substitute for reinforced concrete but as a more appropriate material in particular applications. In some respects, Sutherland's statement reflects views expressed by Dieste on the education of engineers

'*…we shouldn't educate young people by teaching them tables and manuals when they don't know the fundamentals behind them. Instead, we should give them a solid base in the scientific fundamentals of their field and through workshop courses teach them how to use these fundamentals.*'[141]

While Sutherland was rather cautious, Curtin[142] was very bullish about the potential of structural masonry in general and post-tensioned brickwork in particular. By the early 1980s, his practice had designed and built a number of large single storey diaphragm[143] walls which he promoted extensively in the national construction press. He did not take full credit for the development of the diaphragm wall.

'*The idea of the diaphragm is so simple that many other engineers must have discovered the technique and there is some evidence that Victorian engineers may have used it.*'

In fact Dieste used the diaphragm wall as a logical construction solution to the tower behind the altar. It acts as a buttress to the church. At a height of 18·4 m, it is considerably taller than the tallest (10 m) reported by Curtin.[144] Dieste used the diaphragm wall earlier, in 1966, for the church of Our Lady of Lourdes in Montevideo.[145] It was one of the few parts of the project actually completed. A logical development from the diaphragm wall was the application of post-tensioning which Curtin pursued with typical enthusiasm. A headline in *Building Design*[146] in 1986 described the use of post-tensioning in diaphragm walls as 'Breakthrough in brick design' and the article stated that that a 'top engineer claims to have revolutionised the structural qualities of brickwork'. In the context of diaphragm walls, post-tensioning is used to pre-compress the wall and thereby increase its bending resistance. The true innovation of post-tensioned diaphragm walls is as a simple and cost-effective construction technique. In purely structural terms, the principal actions are in fact identical to the already established behaviour of masonry shear walls in multi-storey load-bearing buildings, where the self-weight naturally induces a pre-compression allowing the walls to resist bending due to lateral loads. In most applications,[147] the walls require much lower levels of pre-compression than experienced in the lower storeys of a multi-storey building. In San Pedro, prestressing is used to carry large, permanent loads over long spans, an altogether more significant structural application.[148]

Dieste's designs for San Pedro pre-date and are clearly more advanced than anything that was happening during the brick 'revolution' in the UK, particularly in his use of prestressing. That his work went unnoticed is not surprising. Dieste himself did not look to the developed world in order to import technology and methods. His earlier travels in Europe had convinced him that not all the advances of a fully industrial economy lead to happiness and contentment and confirmed his conviction that the developing countries needed to solve their own problems themselves. That he did not promote his technology abroad to any great extent is probably because he felt that his approach was particular to Uruguay and South America. As Sutherland had said, engineers do not tend to study the built work of others to the same extent as architects

Fig. 5.19. Rosette window.

Fig. 5.18. Rear of church from presbytery.

Chapter Five
In the footsteps of tradition

and instead look to codes of practice and established precedent for security and guidance.[149]

As if leaving the best to the last there is one more remarkable detail in the church. It is perceived by the departing congregation at the end of the mass. The wall over the entrance to the nave is perforated by a large rose window, occupying almost all of the wall (*Fig. 5.18*). The window is not glazed and draws light from windows on the original façade. The large opening is broken into smaller sections using irregular, concentric hexagonal diaphragms (*Figs 5.19 and 5.20*). The sides of the diaphragms lie parallel to the slopes of the nave wall and roof and are relatively deep planes which help smooth the light into a steady glow. The window on the external façade, which is both offset and framed differently, is not visible through the rose window. The congregation only sees the light through a series of disembodied rings floating in space. There are many spiritual and symbolic interpretations that could describe this experience. For example, the host raised by the priest at the moment of transubstantiation, the very essence of the mass itself when the word was made flesh.[150] However, even the confirmed agnostic could not fail to be uplifted in some way. In constructing the window, Dieste completed the renovation to the nave. Dieste was unhappy with the appearance and quality of the original windows and decided to replace them. In this he went beyond the brief for the project was

Fig. 5.20. Detail of rosette window.

Fig. 5.21. Window over presbytery.

originally only to repair the roof. Given Dieste's concern for detail and thoroughness one has the impression that it was inevitable that the original window would be replaced. Dieste followed a logical path in the conception and design of the internal space creating a consistency both in surface and control of light. The original window in the nave wall would have acted in opposition to this. The window was framed by iron columns, supporting the bell tower above. These were removed and replaced by a concrete portal frame which increased the opening considerably, itself a fairly complex piece of underpinning. If Dieste ever allowed himself some fun with structure it is here. The hexagons are constructed from 50 mm thick reinforced brick and appear to be suspended with no physical means of support.

A steel frame was inserted into the opening and anchored into the walls. Slender steel rods attached to this frame are used to suspend the innermost hexagon, fabricated from welded steel angles. Steel angle sections were then attached to the rods for the intermediate hexagons. Supporting a heavy mass from flexible supports could lead to unwanted movements and instability of the framework. The resolution lay, as with so many of his structures, in prestressing. The supporting structure was installed in two parts, slightly shorter than the opening. The inner frame was heated, causing it to expand, and then welded together in place, holding the whole support frame taut against the opening as the steel cooled.

In San Pedro, Dieste has developed an

almost unique typology. Whereas many contemporary designs, Atlantida included, used a large single roof as a unifying device, this was not possible given the predetermined floor plan. Nevertheless the inclined nave walls and the unlit side aisles, combined with the earthiness of the handmade brick, create a space that appears sculpted from the ground itself. The use of brickwork in this particular bonding pattern creates interesting chromatic effects. The ratio of joint thickness to brick is greater than normal and therefore provides a higher degree of reflectivity.[151] There is no direct light penetrating the building. All the light entering is diffused and reflected off the brickwork (*Fig. 5.21*).

Whereas many of his other forms of construction were developed, repeated and refined into established construction systems, Dieste conceived a unique and innovative design which satisfied the particular case of San Pedro.[152]

The work in San Pedro extended beyond the original brief to repair the roof, largely due to Dieste's vision and confidence in his engineering skills. He achieved the technical and structural ambitions that his concept demanded and was justified in terms of cosmic economy, 'to be in accord with the profound order of the world'.[153] Nevertheless the reconstruction, completed in 1971, cost only US $27 000.

Chapter Six
Towers

Chapter Six
Towers

In the dialogue between architecture and engineering, the balance of importance of the disciplines relates to the significance of the structure to the design. There can be structure without architecture but there cannot be architecture without structure. For small buildings, structural needs can be met by traditional construction methods and will not necessarily 'compromise' the architecture. Often the structure is hidden within the fabric. In tall, slender buildings, stability is a dominant factor and the structure and the semiotics of the structure are so apparent as to be inseparable from the architecture.

Humans, unlike most other mammals, are bipeds, with a high centre of gravity and one limb fewer than required for complete stability.[154] The human stance, tall and erect on two feet needs constant, subconscious adjustment of foot position and body weight to remain upright. A highlight in a parent's experience of childhood occurs when the child starts to walk; the complex mechanism of stability has been mastered. Humans therefore become acutely sensitive to stability from a very early age. In comparison with other building types, the tower is the most anthropomorphic, its height being much greater than its breadth, a comparison also valid between humans and other animals, the majority being quadrupeds. In most cultures, being tall is generally seen as a more positive physical characteristic than being short. Towers are therefore able to speak in a language more direct and potent than most other structures. This language has been used to communicate aspiration, spirituality and power.[155] The Leaning Tower of Pisa, next to the Eiffel tower, is surely the most well-known tower in the world and fascinates entirely because of its apparent defiance of a human sense of stability [156] rather than any particular architectural or technical merit of its original design. Each culture has its tower, the spire, the campanile, minaret or pagoda. The symbolism of tall buildings and towers had been analysed by many architects and engineers.

In a comprehensive history of the tower;[157] the highly respected engineer Fritz Leonhardt and the architect Edwin Heinle, ascribe a dual nature to the tower. In many civic, religious and even defensive buildings the symbolism of the tower is more important that its functional duties. In contrast there are certain towers where height is required purely for functional reasons for example, lighthouses, windmills, water towers and transmission masts. Nevertheless these towers cannot escape from the symbiotic relationship demanded between form and structure. Dieste has built many towers, most of which fall into the latter category, where the height is needed for operational requirements. Again, as with the vaults, he has developed an appropriate language which considers carefully the interrelationship of form, structure and construction process while challenging preconceptions about the use of materials. The

Fig. 6.2. Bell tower - Church at Atlantida.

Fig. 6.1. Bell tower - Church at Atlantida.

form has developed and improved with successive application.

An early tower was the campanile of the church at Atlantida. It is relatively modest in scale and constructed entirely in reinforced brick (*Fig. 6.1*). The spiral staircase on the inner circumference is cantilevered brickwork. The tower is independent of the church situated to the side, set back from the front elevation (*Fig. 6.2*). In making this arrangement Dieste allows the forms of both buildings to maintain their own integrity and is also playing a game with the traditional symbolism of the bell tower. In the normal convention, the bell tower is that part of the church which is used to summon the faithful, an obvious sign of the power of the clergy. By separating the tower from the church, he encourages the tower to be something else. There are now many examples of churches with separate bell towers, Dieste is, however, explicit in his reasons:

'Of course a bell tower is built to hold up bells, but it is also built so that young couples can climb it on Sunday to discover the landscape and so that children playing in it can relive stories from long ago that sleep inside each one of us. It is also built to contemplate space, especially in spring when the swallows surround it like live arrows.'[158]

The bell tower is a further challenge to the accepted relationship between the formality of the traditional church building and the relationship with the laity, an element of the church that he hopes will be embraced and possessed by the people in the same way as he broke down the barriers between the altar and the nave, as discussed in chapters four and five.

Water towers are used to maintain a local water supply at an adequate pressure, usually in areas of flat land. Their use expanded and developed with the Industrial Revolution. In the nineteenth century most water towers were constructed of brick, using thick walls for the shaft supporting the tank. The tank containing the water was constructed of riveted steel plate.[159] Many of the early water towers were very large, urban structures intended for domestic supply and fire fighting. Some were beautifully designed and have been preserved past their original effective life as valuable historic monuments. In the twentieth century, steel and concrete took

Chapter Six
Towers

over as the predominant materials for water towers. Some are towers in the broadest sense of the word, sculpted objects of great presence, while other are merely elevated tanks.

Leonhardt and Heinle are scathing in their criticism of the latter:

'There are many ugly water towers in the world, usually they were designed by insensitive engineers, who in many instances were not the best representatives of their profession.'[160]

In other words; structure without architecture.

Dieste's towers derive their beauty from the careful and sensitive geometry of simple forms. He has built a number of water towers, using the same techniques as the bell tower at Atlantida (*Fig. 6.3*). Typically, the shaft of the tower is formed as a gently tapering, truncated cone of brick masonry (*Fig. 6.4*). The sides of the shaft are perforated with large vertical slots. Each column of slots is staggered vertically in relation to its adjacent column (*Fig. 6.5*). The size and position of the slots has been carefully considered and they serve a number of functions. They provide a degree of permeability that reduces the wind forces on the shaft. The change in the shaft diameter as the tower rises (*Fig. 6.6*) is facilitated by reducing the width of the slots, allowing the continuous vertical ribs of brickwork to remain constant in width and therefore easier to construct. During construction the slots provide a simple support for a working platform on both the inside and outside of the shaft (*Fig. 6.7*).

Fig. 6.3. Water tower - 60 m³.

Fig. 6.4. Water tower at Refresco del Norte.

Fig. 6.6. Shaft of water tower.

Remo Pedreschi

Dieste Archive

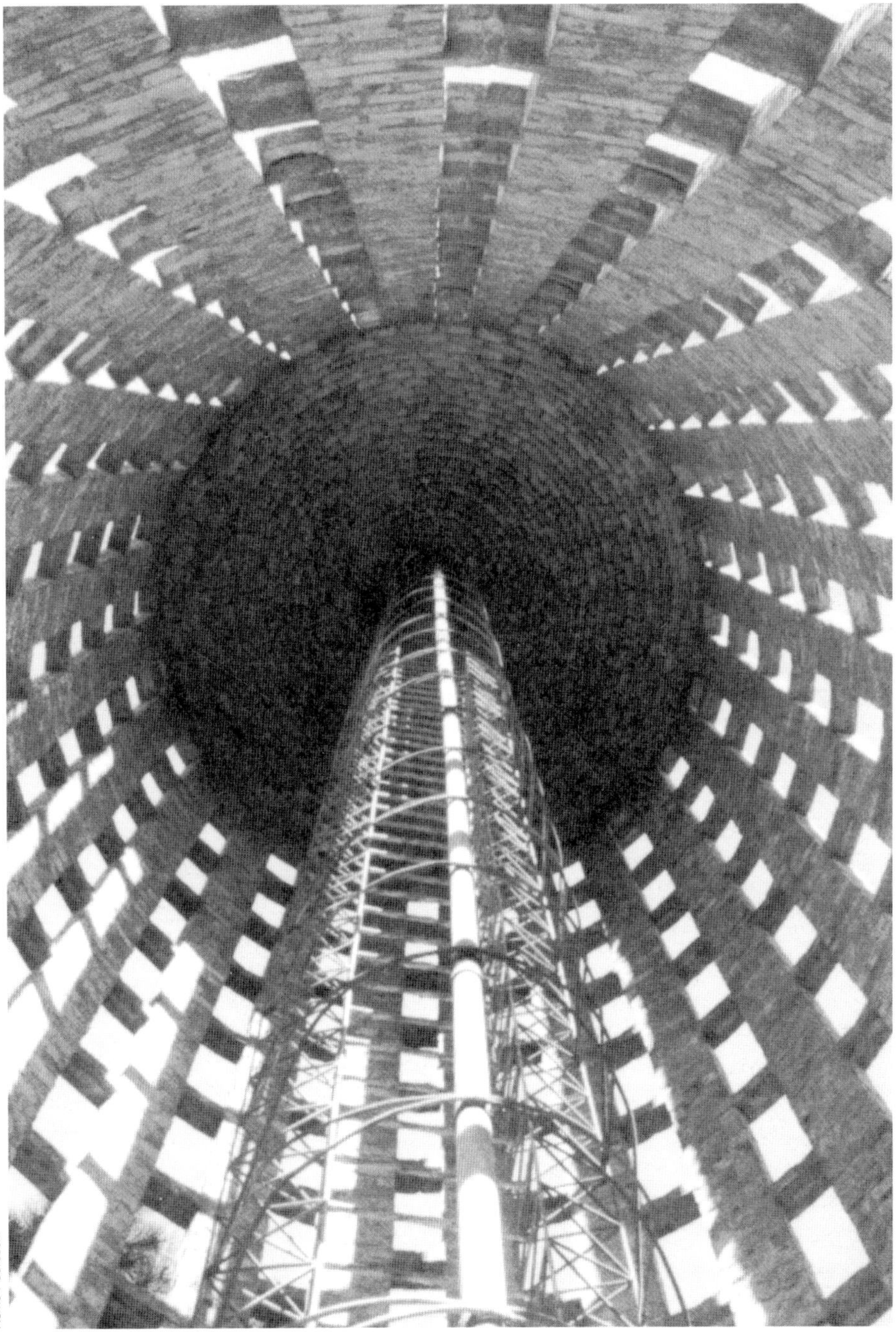

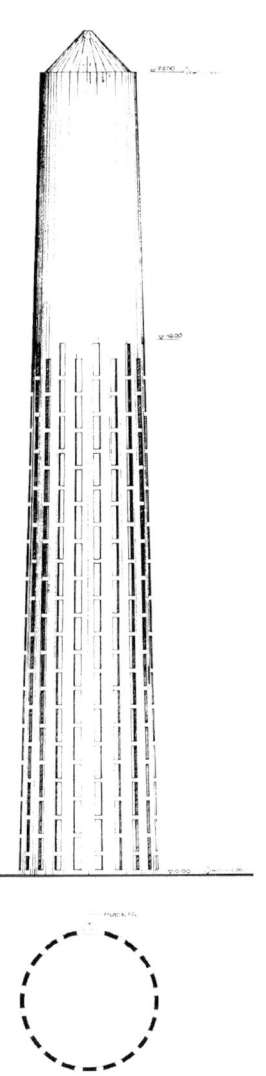

Fig. 6.5. Water tower.

Chapter Six
Towers

Therefore scaffolding is not needed, a major consideration in a tall structure. The shaft is set out simply by taking great care to establish the correct inclination of the taper for the first four or five metres of the shaft then continuing to follow this line to the top using simple straight edges. It quickly becomes clear that brick is the ideal material for this structure, simpler to use than concrete as formwork and day-work joints are eliminated and the finished quality of the surface is more readily assured.[161] The vertical ribs of the shaft are connected by intermittent horizontal ribs of reinforced brickwork, creating the staggered pattern of the slots. Staggering the ribs avoids the appearance of continuous bands around the shaft, which tend to make the shaft appear broader and less slender.

The structural behaviour is straight forward. The shaft is designed to act as a circular column (*Fig. 6.8*). The wind loads are based on a three second gust[162] and are assumed to apply a static horizontal force to the tower.[163] The axial tension and compression shaft due to the

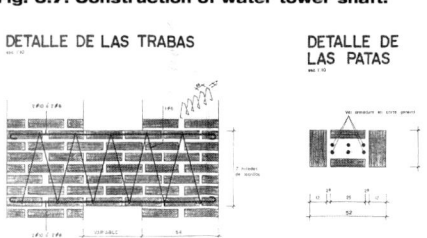

Fig. 6.7. Construction of water tower shaft.

Fig. 6.8. Section through water tower.

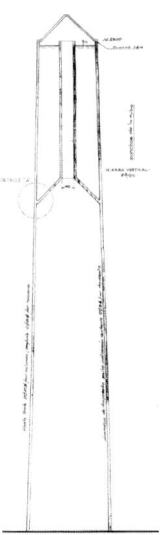

horizontal and vertical forces are carried by the vertical ribs. The tensile forces are carried by steel reinforcement running through the ribs. The cross ribs tie the vertical ribs together, allowing the shaft to act as a complete section. The cross ribs are reinforced to allow shear transfer between the ribs.[164]

The water tower shown has a capacity of 60 000 litres with a minimum head of water of 16 m above ground level. The inclination of the shaft continues into the tank itself. The tank is terminated by a conical roof, giving the overall appearance of a large brick Egyptian obelisk, over 25 m high. In the brick water towers of the nineteenth century the tank itself was normally constructed from riveted steel plates, the

Fig. 6.9. Detail through tank wall.

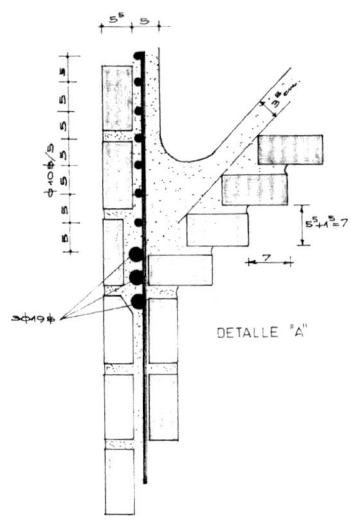

brickwork used only to form the vertical shaft and to clad the tank. In this tower the tank is also constructed using reinforced brickwork. The base of the tank is constructed as a truncated cone of corbelled brickwork. A series of circular brick courses are stepped progressively, one on top of the other, until they meet the internal shaft of the tower. Under the action of water pressure (total head of 8 m), the cone will compress in on itself, providing that the lateral thrust at the base of the cone is contained. The section at the base of the cone is heavily reinforced with hooped steel bars. In any water retaining structures flat surfaces resisting water pressure experience very large bending moments which need thick, heavily reinforced slabs. In this example, the base of the

V. Del Amo

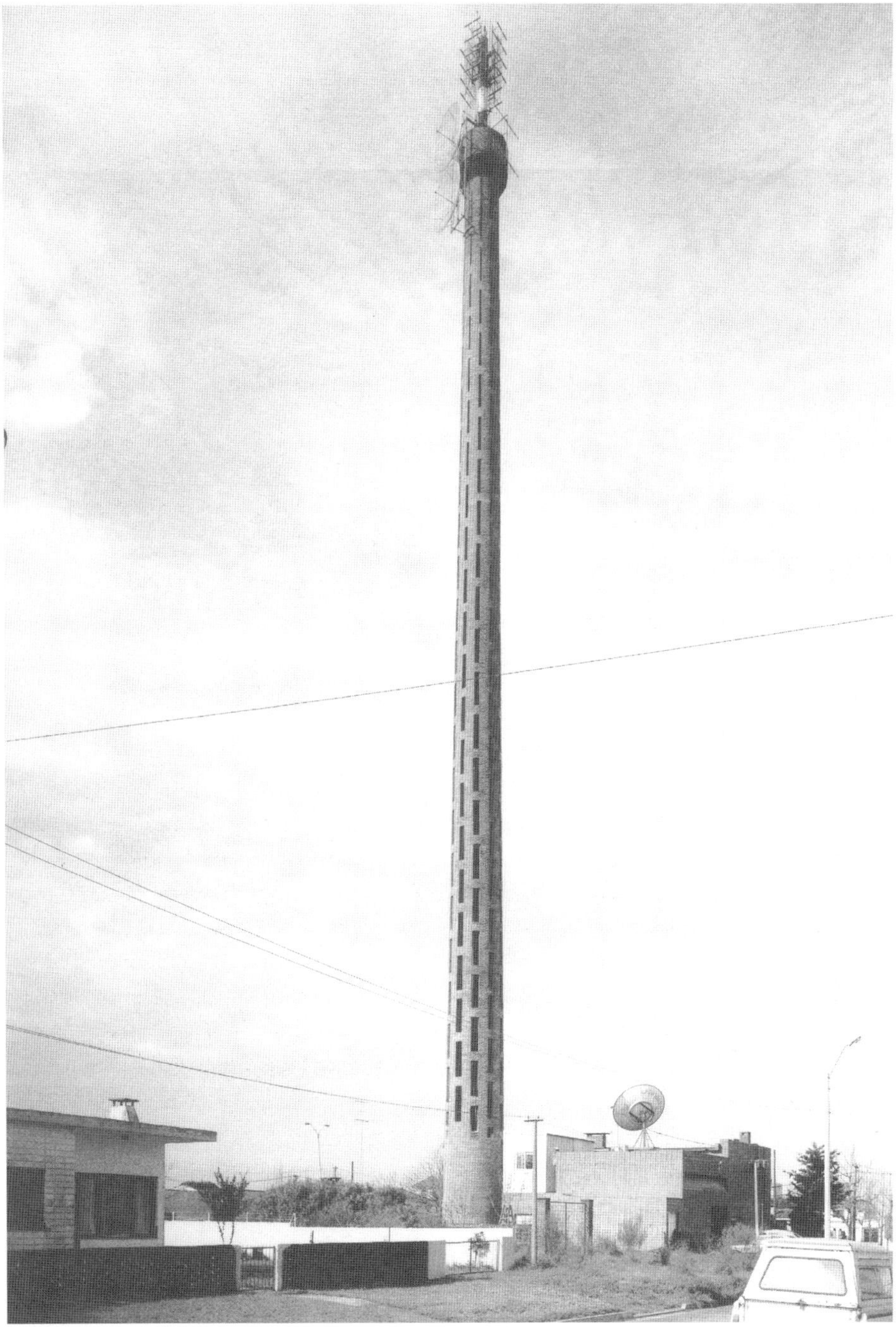

Fig. 6.10. TV communications tower - Maldonado.

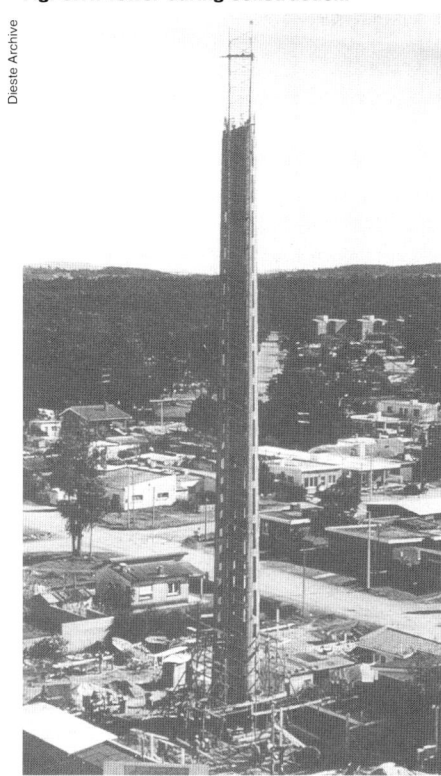

Fig. 6.11. Tower during construction.

Dieste Archive

tank is only 6·5 cm thick, comprising one course of corbelled brickwork and a 3·5 cm thick concrete render. The external wall of the tank (*Fig. 6.9*) consists of a single layer of bricks reinforced using 10 mm diameter hooped steel bars at 5 cm spacing, lined internally with 5 cm of concrete render. The water tower for Fagar Cola is different from other towers in that the shaft of the tank is solid. It is almost a pure cylinder of brick, with only a very slight taper in diameter from 4·25 m to 3·6 m over a height of 26 m.

The communication tower developed an important iconic status in the latter half of twentieth century. It is the symbol of the 'global village'. Dieste's construction methods for towers

Chapter Six
Towers

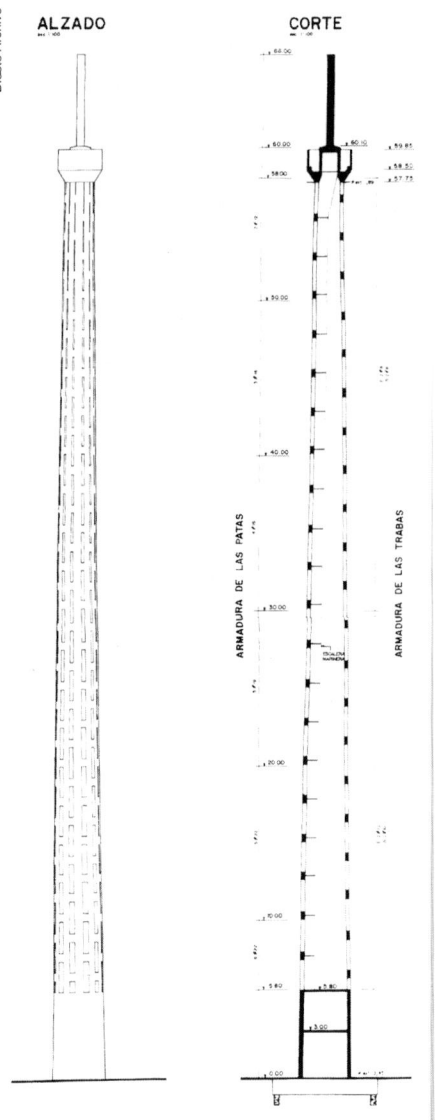

Fig. 6.12. Elevation and section of the TV tower.

were stretched considerably in 1986 with the completion of a 66 m high television communications tower in Maldonado, north of Montevideo on the Atlantic coast (*Figs 6.10 and 6.11*). The tower itself is within one kilometre of the ocean. The shaft takes the same form as the water towers, but it is of a much more slender construction (*Figs 6.12 and 6.13*). The diameter at the base is only 3·5 m and tapers to less than 2 m at the top of the shaft. On top of the shaft sits a cylindrical tower head to support the antenna mast and various parabolic directional transmitters (*Fig. 6.14*).[165] The tower head projects from the shaft on corbelled brickwork. For formal reasons, Dieste would rather that the head did not project, preferring to maintain the geometry of the truncated cone. The shape of the head was dictated by the installation requirements of the telecommunication engineers. The wall of the shaft contains the vertical reinforcement within a grouted cavity. The cross ribs that connect the vertical columns of brick are much deeper than the lower water towers, as it is necessary to increase the rigidity of the shaft. The foundations of the tower consist of a circular concrete slab supported on a ring of bored piles (*Figs 6.15 and 6.16*).

The tower at Maldonado echoes some of the ideas proposed by Leonhardt in favour of the use of reinforced concrete over steel in communication towers. Namely, that concrete is better than steel in terms of cost, rigidity and damping characteristics, all attributes of brickwork, and he further concluded that the ideal geometry is that of the truncated cone for minimal wind resistance. Brickwork would meet his requirements equally well and is probably cheaper than concrete. Leonhardt has built a number of such communication towers. Many incorporate additional functions such as restaurants and viewing platforms so a direct comparison is difficult. The closest comparable tower in concrete is the Emley Moore mast in England by Ove Arup and Partners. The tower is 330 m high and has a base diameter of 24·40 m tapering to 6·5 m at the top. Using ratio of height to base diameter provides an approximate comparison of the relative slenderness, 13·5 for Emley Moore and 18·8 for Maldonado, Maldonado is considerably more slender.[166]

A number of towers have been commissioned to celebrate events of international importance. They aim to be striking in appearance, such as the Collserola tower for the 1992 Olympic games in Barcelona by Norman Foster or to create landmarks for major cities such as the CN tower in Toronto, the world's largest tower at 553 m. In many towers such as these there is considerable innovation, often necessary to achieve architectural rather than operational needs. However, it is the tower at Maldonado that is singled out in the introduction to the book, *Communication Towers*.[167]

'Unique design and architectural value are not only found in association with colossal height. This is demonstrated by the moderately high Maldonado

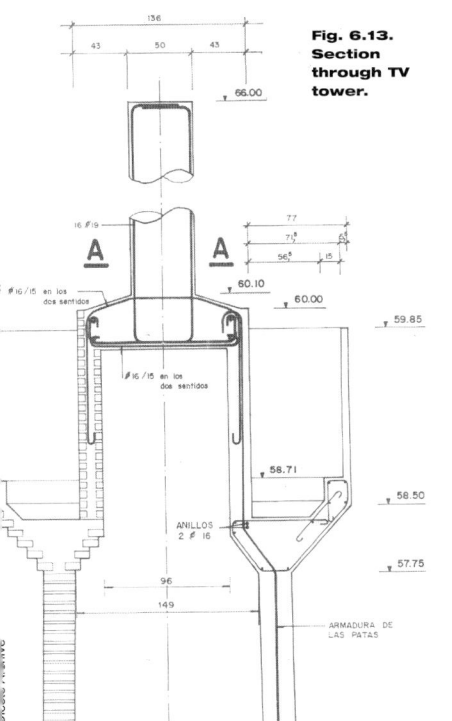

Fig. 6.13. Section through TV tower.

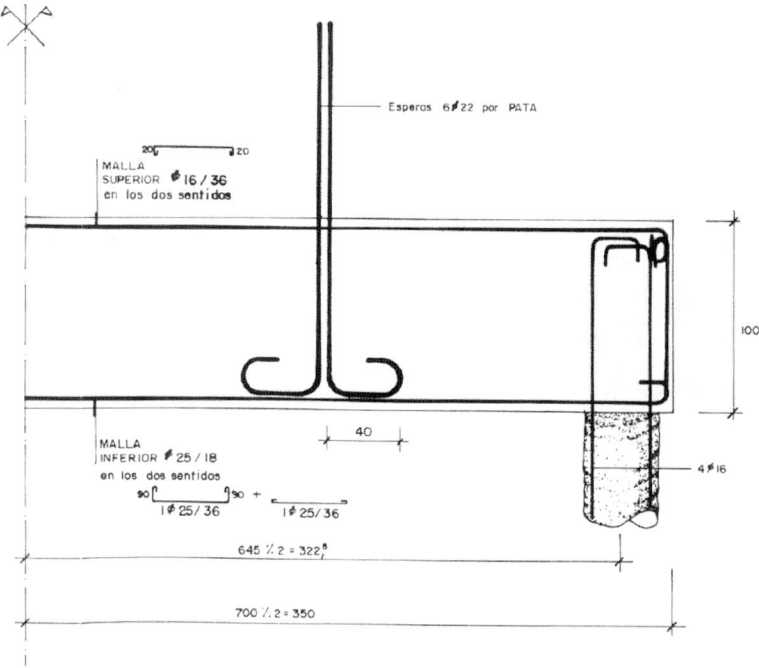

Fig. 6.15. Foundation of TV tower.

Television tower by Eladio Dieste.'

The towers are interesting objects in the repertoire of Dieste's experiments with structural masonry. In their form they are much closer to towers in reinforced concrete than the earlier brick towers, being similar in both slenderness and thickness. However, they are the products of a refined design that demonstrates the potential of this 'new material'.

Fig. 6.14. Top of tower.

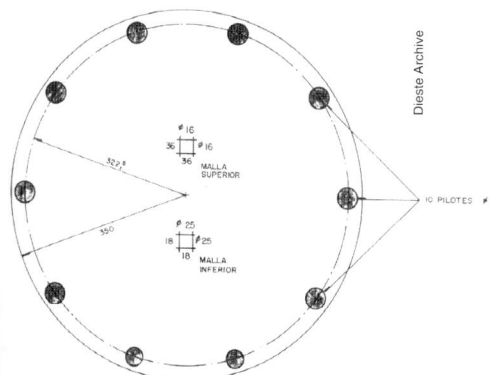

Fig. 6.16. Arrangement of pile foundation - TV tower.

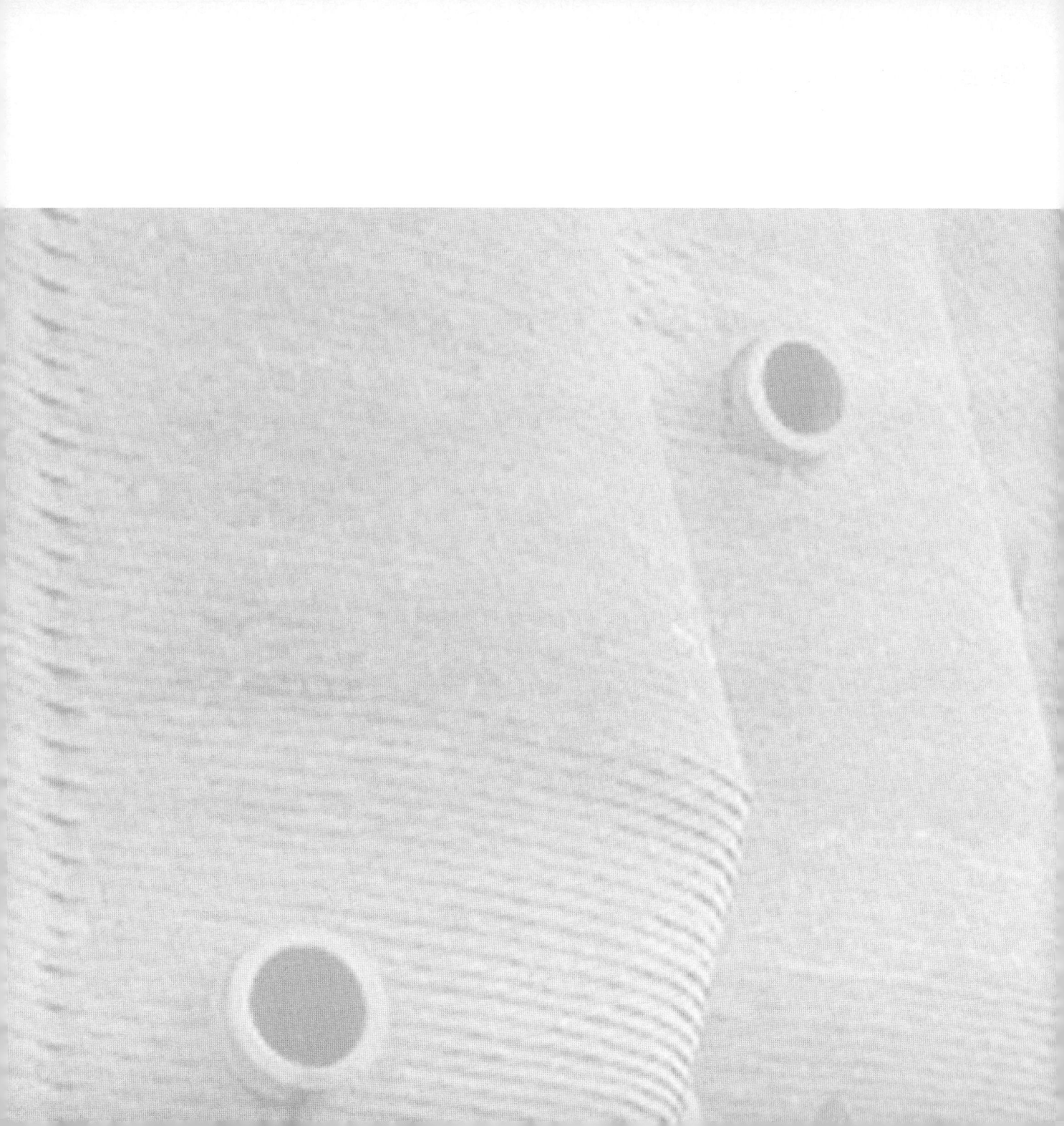

Chapter Seven
Typological Variations

Chapter Seven
Typological Variations

Dieste Y Montañez have undertaken many projects. The preceding chapters have presented characteristic examples, extracts from their portfolio of buildings. Their work could be subdivided into two broad categories, the generic and the particular. The free-standing barrel vault, the Gaussian vault and the tower are all examples of the generic. These forms have been used repeatedly, developed and refined, to become building systems where the costs and the construction programme can be readily determined from prior experience. The methods and technology of reinforced brick is highly flexible. The two churches, at Atlantida and Durazno, are examples of the particular. The design and construction of these buildings were carefully considered and executed and the results greatly individualised. In both cases Dieste reinterpreted and extended the original brief to expand their architectural and social potential. There are other examples of his construction techniques and design philosophy being tailored to particular architectural and cultural challenges. This chapter describes some of the projects that illustrate different aspects of his work or extend the application of his ideas about structural form and architecture into different functional typologies. The projects considered include a large shopping mall, a house, a restaurant and an unfinished church in Montevideo.

Montevideo Shopping

Montevideo Shopping is a large retail development in Montevideo (*Fig. 7.1*). The shopping mall is a large scale building type that expanded rapidly in the mid-to late twentieth century. The growth and development of the mall has been extensive over the last 30 years.[168] Malls

Fig. 7.1. Montevideo Shopping.

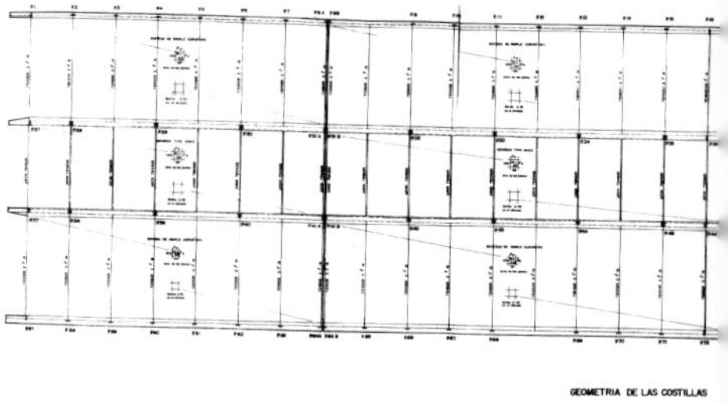

Fig. 7.2. Montevideo Shopping - plan of main floor.

signify the growth in shopping as a leisure activity, a place to be as well as to use, to be enjoyed by a mass population. They are very much a product of the developed world, a consequence of a consumer-led economy. Originally, malls were seen primarily as indoor streets and were designed to focus the shopper's attention on the store fronts. In more recent designs, greater attention has been placed on the mall itself, the aim being to create a consumer friendly environment, a desirable place in itself. By a combination of subtle touches and bold initiatives customers are enticed. As a type the mall is a carefully planned and organised building, where the correct blend of economics, demographics, marketing and finance is critical for commercial success. Typically, emphasis is placed on the procurement and design of the building; in speed of construction,[169] in value for money and in a high quality, both in the finishes and the internal space.

Dieste y Montañez completed Montevideo Shopping in 1985, one of the first malls in the city. The mall is situated on a sloping site near the centre of the city. The project was a great challenge to the firm to use their methods in such a contemporary building type within the strict programmatic requirements and time constraints. Could their apparently 'homespun' technology compete successfully with the steel, glass and concrete buildings that characteristically defines the current mall. In their other projects, the construction programmes were fast, due largely to improvements in their construction techniques that had been developed for efficiency of the methods themselves, rather than specifically for client pressure. In projects, like the churches, often working to an agenda that Dieste was instrumental in developing, the rate of construction was slow, to allow careful treatment of details and finishes. In the mall there was pressure for early completion and the expectation of a high quality of workmanship and finish from the client.

The company worked closely with the architects, Goméz Platero-López Rey, who led the design team and were responsible for overall planning and design. Dieste's company was responsible for the design and construction of all structural aspects of the project. Even with Dieste acting as second consultant, the building is evidently Diestian in character and indeed is a catalogue of his innovations, including barrel vaults, Gaussian vaults, curved walls and a water tower.

The main building has a total floor area of 10 000 m2[170] on two floors. The plan is rectangular, nominally 120 m long and 40 m wide

Fig. 7.3. Montevideo Shopping – central arcade.

(*Fig. 7.2*). The site steps down a full storey between the two long elevations of the building with car parking at both upper and lower ground levels. The main entrances to the mall are from the car park at the upper ground level. The supermarket and shops are contained on the upper storey in two 16 m wide strips to either side of a top-lit arcade. The shops sell a mixture of quality clothing and consumer products. The lower floor contains the ubiquitous open plan food court as well as storage and services for the shops and supermarket. The roof plan follows the functional layout of the building. Two barrel vaults, each spanning 16 m, cover the shopping zones while the central arcade (*Fig. 7.3*) is covered with an 8 m span Gaussian vault (*Fig. 7.4*). The Gaussian vault is used for architectural rather than

Fig. 7.4. Montevideo Shopping – Gaussian vault

Chapter Seven
Typological Variations

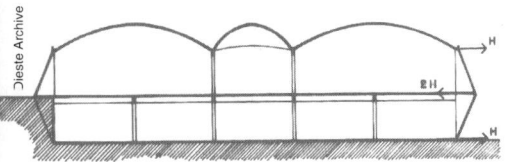

Fig. 7.5. Montevideo Shopping - section.

Fig. 7.6. Montevideo Shopping - window in central arcade.

Fig. 7.7. Montevideo Shopping - side walls.

structural reasons and becomes the central focus of the mall itself. The three vaults sit on their springing points at the same level and rise to their apex at the same height (*Fig. 7.5*). The Gaussian vault over the mall, having a shorter span than the other two vaults, therefore has an exaggerated geometry,[171] with much greater curvature than other examples, such as the warehouse at Montevideo Docks.[172] The extenuated curvature of the central vault generates deep, curved windows between the vaults producing a brightly lit arcade (*Fig. 7.6*). The pronounced curvatures of the underside of the central vault progresses along the mall in a series of waves.[173] Due to their exaggerated vertical scale, the vaults appear as if they have undergone a great expansion, causing them to swell upwards and finally burst, letting

the light cascade inside. The patterning of the bricks on the underside of the vault adds to the dynamics of the surface; the tight curvatures of the surface result in considerable dislocations of the bonding pattern between the blocks as they are adjusted to suit the curved form of the vault.[174] The larger barrel vaults over the shops have a simpler geometry and a more regular pattern of bonding.

In common with many shopping malls Montevideo Shopping is inward-looking, the central arcade bounded by the shops, the fenestration in the external walls restricted. The architects were concerned that these long walls might appear monotonous and bland and wanted the wall to undulate in some manner, particularly at the lower end of the site where the wall is

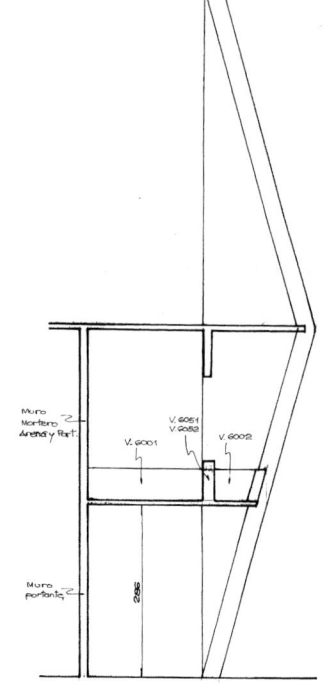

Fig. 7.8. Montevideo Shopping - section through wall.

11·5 m in height. Dieste created a form for the walls that undulated and at the same time provided a structural function (*Figs 7.7 and 7.8*). At ground level, the walls are straight and undulate as they rise, similar to the church at Atlantida (*Fig. 7.9*). At mid-height the undulations are at a maximum and then decrease in amplitude to finish again as a straight edge at the eaves (*Fig. 7.10*). The sequence of large, curved ridges appears like the work of an angry giant, who has hammered from the inside with a large clenched fist. The deformations echo the warped plastic character of the central vaults (*Fig. 7.11*). The wall has two structural functions; to support the intermediate floor and to buttress the horizontal thrusts of the vaults. The intermediate floor, the shopping floor, connects with the wall at the apex of the ridges of the wall, (i.e. at mid-height) and continues across the building to the upper end of the site where it coincides with ground level. On this elevation, only the top half of the wall is visible, sitting on a ground beam connected to a ribbed retaining wall that extends to the basement floor level.

The surface of the wall is defined in a similar manner to the church at Atlantida. A series of vertical straight lines, starting from a straight horizontal line at the base of the wall is displaced to follow a horizontal sinusoidal curve at the mid-point of the wall, returning to a straight horizontal line at the top of the wall (*Fig. 7.12*).[175] The walls are tied together using prestressed steel tendons, passing through the floor slab, connected on one elevation to the top of the retaining wall and on the other to the mid-point of the opposite wall, passed through floor slabs. There are relatively few windows on the walls, emphasising the inward looking character of the building. Small, circular concrete windows pierce the walls, like deep portholes, in a regular pattern. They are perhaps, slightly uncomfortable and perhaps unnecessary intrusions [176] on the rippling surface, particularly when compared with the small rectangular windows in the walls of the church at Atlantida, designed to minimise their impact on the surface of the walls.

Fig. 7.9. Montevideo Shopping – external wall.

The structure of the building is an unusual hybrid of reinforced and prestressed masonry and conventional reinforced concrete. The internal frame, supporting the slab and main floor level, is constructed using an in situ concrete frame and slab on a square grid of 8 m. The shops occupy two bays on either side of the single bay of the arcade. Two lines of columns extend above this level, on either side of the central mall and support continuous concrete beams running the full length of the building. The beams provide both the vertical support and the interface between the barrel vaults and the Gaussian vault. The other ends of the barrel vaults are connected to a concrete edge beam supported on the brick side walls. The roof gutter is contained behind an upstand to the edge beam. The external walls have an overall thickness of 300 mm that consists of two skins of brick, the cavity includes a 20 mm layer of polyurethane insulation.

Chapter Seven
Typological Variations

Fig. 7.10. Montevideo Shopping - roof and wall.

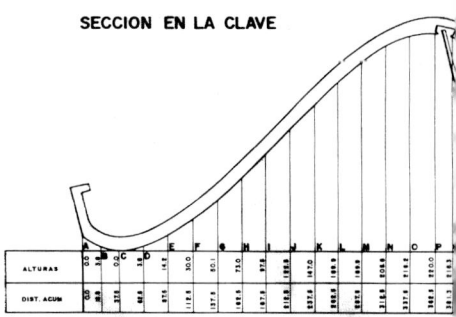

Fig. 7.11. Montevideo Shopping - section through central vault.

The thrust from the roof vault is resisted by the external walls, which act as vertical cantilevering slabs (*Fig. 7.13*). The prestressed cable at the level of the main floor ties the side walls together. The wall bends about the slab at mid-height. The maximum bending moment due to the thrust from the wall occurs at this point, where the undulations are greatest. The vertical profile of the wall therefore has the same shape as the bending moment diagram generated by the roof thrusts. The cables are anchored into the buttresses at ground level at the top end of the site and stressed from the opposite end. The anchorages at the stressing end[177] of the prestressed cables are contained within concrete anchor blocks built into the wall and covered externally with brick slips (*Fig 7.15*). On the

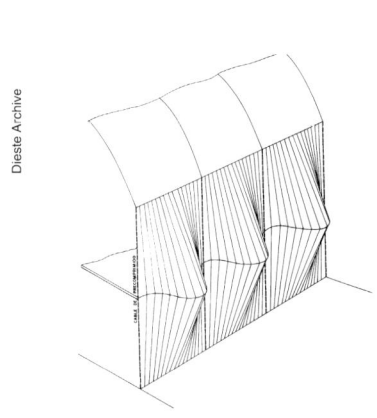

Fig. 7.12. Montevideo Shopping - geometry of wall.

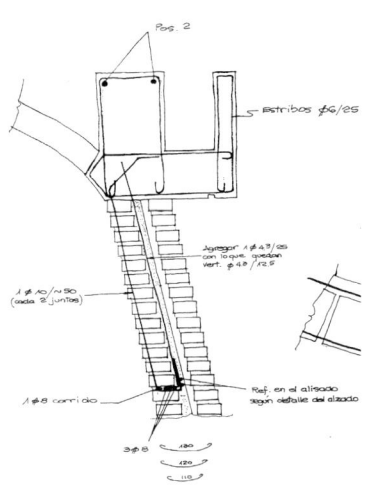

Fig. 7.13. Montevideo Shopping - roof wall junction.

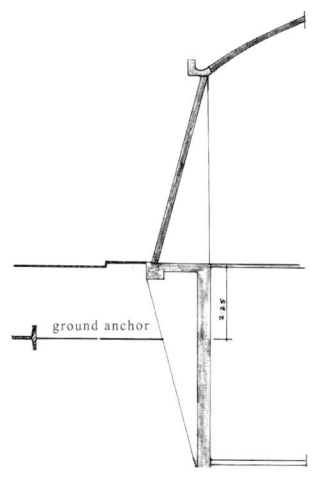

Fig. 7.14. Montevideo Shopping - retaining wall.

opposite side of the wall, the cable is anchored to the top of the retaining wall. The walls are also prestressed vertically to counter the tensile stresses induced by bending. Cables are placed within the wall between the undulations, and are anchored to the foundations and the edge beam at the eaves.

At the upper end of the site, the wall itself is has an interesting design. Above ground, the wall curves in the same form as the opposite wall from mid-height upwards. Below ground, the wall is supported on a 6 m deep retaining wall (*Fig. 7.14*). The retaining wall has a flat internal surface with projecting ribs that taper from the ground level downwards, a reversal of the normal direction for the buttresses of such walls.[178] The ribs coincide with the undulations of the external wall and support the ground beam on which the curved wall is constructed, tying the wall to the main structure and avoiding further foundations. The lateral pressure from the soil and car park above on the retaining wall is restrained using a series of ground anchors which are attached to the ribs at a depth below the ground coincident with vertical centroid of the ribs. The anchor consists of a reinforced concrete T-beam lying on its side, 2·3 m in length. Four, 22 mm diameter steel tension cables are used between each anchor and each rib. The unconventional wall was a solution that no doubt drew on Dieste's earlier experience with the foundation engineers, Veirmond, that satisfied the conflicting requirements for the curved foundations of an external wall with the need for a buttressed retaining wall. Inverting the taper of the buttress, in some ways counter intuitive, simplifies

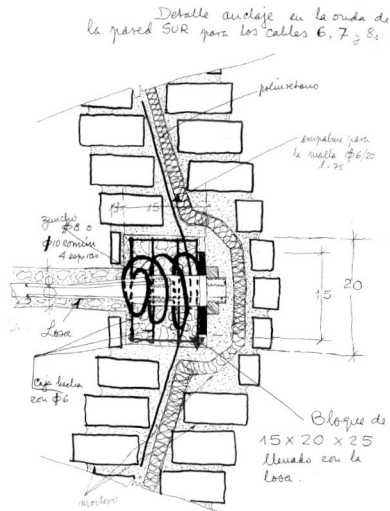

Fig. 7.15. Montevideo Shopping - detail of tension cable.

Chapter Seven
Typological Variations

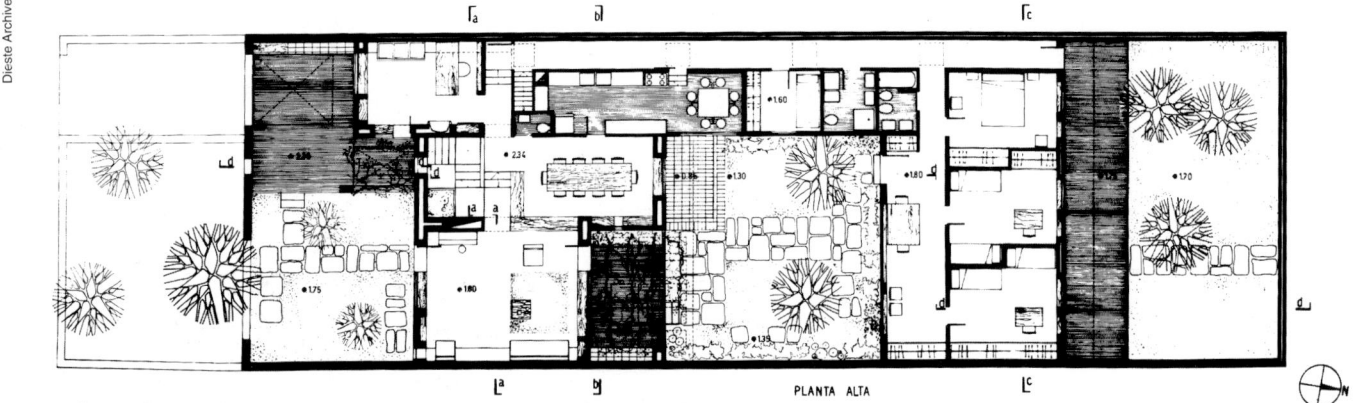

Fig. 7.16. Casa Dieste - floor plan.

the interface between the walls above and below ground and responds to the forces produced by the thrust of the roof.

The earthworks, the walls and the interfaces between the walls and the reinforced concrete frame were novel and particular to the project, while the construction of the roof vaults were now standard procedures of the firm following the methodology and practice described in earlier chapters and representing possibly the least problematic part of the undertaking.

The construction programme was very tight, starting in November 1983 and finishing in April 1985, and at its peak employing a labour force of 350. The project required a considerable amount of management on site which was entrusted to Vito Pacheco, one of the company's long serving and trusted foreman. Pacheco was chosen by Dieste on account of his intelligence and sharp wits which came in useful on a number of occasions.[179]

Montevideo Shopping is an example of a contemporary building type with origins in the developed world, constructed using an indigenous, traditional material in a technologically advanced manner following modern construction management principles, more usually associated with steel or concrete.

Casa Dieste

Dieste's earliest experiments with brick vaults were in the construction of houses, working with the architect, Antonio Bonet, on the Berlinghieri house. He returned to the use of brick vaults in housing some 20 years later with the construction of his own house in 1968. The house was designed for his large household, comprising his wife, their 11 children and a housemaid. By this time, he had extensive experience in the design and construction of brick vaults and the guiding criteria for the house were architectural rather than structural. Not surprisingly, there was a strong concern for how the building should be used and how it would meet the needs of the family. In his own writings, he has described these as the dominating considerations in the design, with a concern to balance both the requirements of a communal family life with the need of the individual for personal space.

His own description of the project expresses some of the ideas developed by Christopher Alexander, subsequently in his book, *The timeless way of building*.

'*Those of us concerned with buildings tend to forget that all the life and soul of a place and all of our experiences there depend not simply on the physical environment but on the pattern of events which we experience there.*'[180]

Dieste wanted to ensure that:

'*...apart from the natural landscape and the landscape that the architecture creates, the human* landscape is also magnificent. In other words to ensure that those that live in the house see each other within the different interior spaces.'[181]

The criteria used in developing the house were:

- to maintain privacy and intimacy
- for all rooms to have access to direct natural light[182]
- to have views to the sea
- multiply the living areas to provide contact and at the same time independence from each other
- to create a sense of spaciousness and nobility
- to control the penetration of light. Dieste has criticised the excessive use of glass in buildings which is both expensive and often requires additional solar or thermal control measures. According to Dieste glass is often used inappropriately.
- '... confusing the unlimited with the infinite'[183]
- to use natural methods for environmental control.

Alexander argues that the best buildings evolve from understanding how people utilise and inhabit space; that buildings are a collection of 'patterns'[184] which themselves have evolved from the interaction of people and buildings. He argues further that the essence of this understanding is the withdrawal of the ego from the process of design.

'But once a person can relax and let the forces of the situation act through him as if he were a medium, then he sees that the language, with very little help, is able to do almost all the work; the building shapes itself'.

This statement helps summarise the approach that Dieste has taken. Having established his own conditions for the building, Dieste follows them with the rigour and discipline of the engineer working almost mathematically. A solution is only acceptable if all his criteria are met. At the time of building his house, the church at Atlantida was nearly ten years old, his firm had completed many projects and he was, by Uruguayan standards, a noted and successful designer and builder. However, a consequence of his focus on the predetermined criteria rather than on the exploitation of an opportunity to show his skill as a designer make his house modest in size and understated in appearance.[185]

The house is situated in the suburb of Punta Gorda to the east of Montevideo on a plot 12 m by 50 m (Fig. 7.16). The site has an elevated position on a north westerly axis and slopes down

Fig. 7.17. Casa Dieste - front elevation.

Fig. 7.18. Casa Dieste - roof terrace.

Chapter Seven
Typological Variations

Fig. 7.19. Casa Dieste - dining and living areas.

10º from back to front. The front elevation to the house faces the sea across the street (*Fig. 7.17*). The house presents a rather austere face to the street, comprising the entrance to the garage and a brick retaining wall at ground level. Above, the wall continues to a second storey, forming a screen to the elevated terrace behind, punctured by two large rectangular openings. From this terrace there are good views, over the street, to the sea, framed by these openings while maintaining the privacy of the terrace from the street (*Fig. 7.18*). The front door of the house is positioned underneath the terrace and is barely visible from the street. Entry to the house is by a staircase from the vestibule, up to the principal floor of the house, the piano nobile, opening directly into the central living space, which consists of separate but interconnected areas of living room, dining room and study (*Fig. 7.19*). The spaces are separated by small changes in level and waist high walls. These living areas open at front and back onto the terrace and an enclosed garden on either side. Virtually all the accommodation is at this level which continues through to the back of the site. The section (*Fig. 7.20*) reveals that the building is constructed as three connected vaults, the soffits of which add further definition to the separate spaces underneath, dividing the dining and living areas. Towards, the rear of the site on the opposite side of the garden, is the bedroom wing consisting of three bedrooms and a wide gallery adjacent to the court-yard. This arrangement allows each bedroom a northerly elevation (*Fig.7.21*).

There are clearly some similarities with the plan of the Berlinghieri[186] house, of some 20 years earlier. Both make similar use of the sloping terrain; both have separate wings for bedroom and living space, and both use brick vaults for the roofs. There are also important differences. In the Berlinghieri house, the bedroom wing lies parallel with the main elevation of the living space and has large areas of glazing presenting a wide open façade to the sea. It is obviously a 'beach house'. In Casa Dieste, the bedroom wing faces the living areas across the courtyard. The building is inward looking, the only external windows around the whole perimeter of the building are the openings in the wall to the terrace at the front. This is a suburban house designed by a father protective of his family. The kitchen and bathrooms are contained in a strip of the building that connects the bedrooms to the living areas and forms a third elevation to the enclosed garden. Both the enclosed garden and the terrace are integral parts of the house, defined by the continuity of the side walls. The roof vault over the living room extends out into the courtyard garden where it transforms into a pergola consisting of an open lattice of brick arches (*Fig. 7.22*). A deciduous vine grows through the pergola creating shade in the summer and permitting light in the winter. The house extends over the greater part of the site, creating the many spaces and differentiated areas that Dieste considered important for his large family. A sense of equality of space and light can be seen in the plan. All the bedrooms are the same size, with the exception of the maid's room, the only room

Fig. 7.20. Casa Dieste - sections.

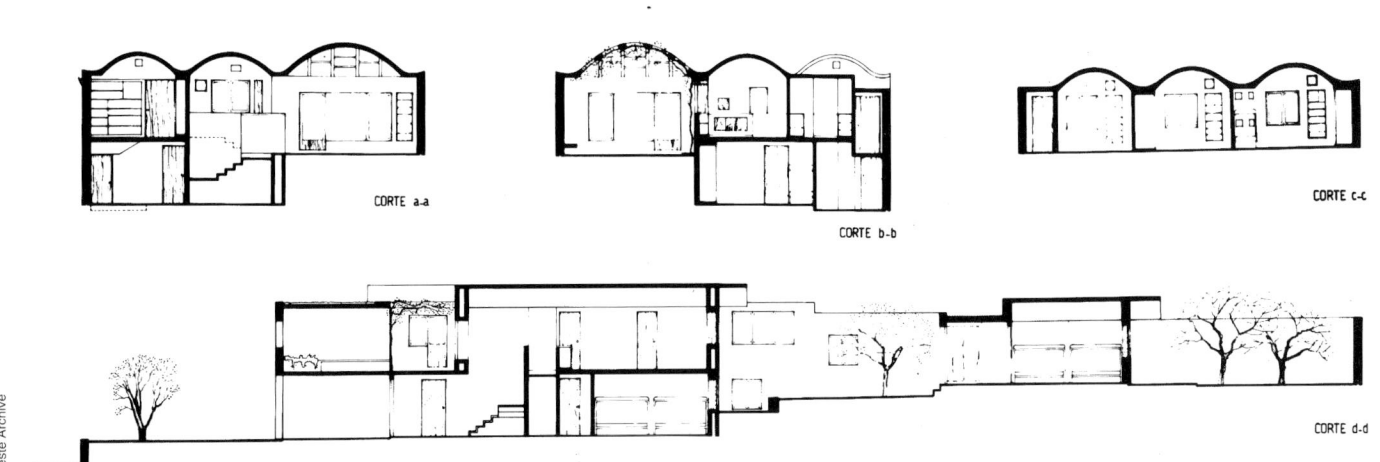

devoted to one person. All the main rooms, with the exception of the study, enjoy good light, even though they are arranged around the three sides of the courtyard (*Fig. 7.23*).

In analysing the needs of his family, Dieste took account of the ages and sex of his children and concluded that he needed four bedrooms, excluding the maid's room. The constraints of the site made it impossible to incorporate a fourth bedroom at the same level as the others without losing access to natural light. Adding an additional storey would have changed so many other aspects of the building, most significantly overshadowing the garden. The solution was to place the fourth bedroom at a lower level, below the dining room, opening it on to a sunken patio in the garden to allow penetration of light.

Fig. 7.21. Casa Dieste - bedroom.

Fig. 7.22. Casa Dieste - enclosed courtyard.

Chapter Seven
Typological Variations

Internally, the house is finished with an almost monastic expression, the finishes are very simple - unplastered brick walls painted white - the soffit of the brick vault is exposed and dark timber is used for window cills, shelves and bookcases throughout. The walls were made thicker than necessary, up to 500 mm thick and incorporate recessed cupboards and book shelves.

The character of Dieste is stamped on the house. He is a serious but modest man. There is nothing ostentatious in the building. It does not announce itself to the world and make a bold statement about the designer. Instead it is a comfortable and secure house. His design skills have been devoted to the meticulous resolution of the plan. It is no larger than it needs to be. There are no guest rooms for example. It could be argued that given the size of his family, a second floor could be justified without being ostentatious. Making the house larger would no doubt have conflicted with his moral sense of economy. The construction of the building requires effort and material that has to be amortised over the life of the house. The size of the family living at home will change more rapidly than the life of the house and will reduce as children get older.[187] Therefore more rooms would more quickly become empty rooms and the house would soon lose its sense of intimacy as it gathered redundant space and fabric. The house is inherently sustainable. This point is demonstrated further in his attitude to the use of glass.[188]

Fig. 7.23. Casa Dieste - window to the sea.

Fig. 7.24. Parador Ayui.

'To open the rooms to the outside but with moderation. I believe that we tend to abuse glass as a building material. If we keep in our minds material needs and their rational fulfilment I don't believe that excessive use of glass can be justified in any climate and much less in ours which is quite extreme.'

Not surprisingly, this view is not only a technical issue of building physics but also an emotional one.

'...in houses that exaggerate the number and size of their windows, we often lose the primitive sense of shelter that a house should give.'

Parador Ayui

The petrol station for Barbieri y Leggire described in chapter two was a manipulation of the construction techniques for barrel vaults to create a structural form both graceful and mysterious, where the weight of the material contradicts the minimalism of its vertical support. Likewise in the design of the roof for the restaurant of the Ayui Inn, also in Salto, Dieste manipulated the form of a traditional heavyweight construction, using an equally clever piece of structural engineering, to create an open, light, summer pavilion (*Fig. 7.24*).

The roof is conical in shape with a base diameter of 15 m, truncated towards the top and supported on 24 slender steel columns. A canopy is formed by a circular horizontal edge slab, cantilevering 3 m from the base of the cone in reinforced brick. The roof is constructed as a brick vault. The language of such a construction normally would include a ring beam at the base of the cone sitting on lintels and piers, a heavy supporting structure to carry a heavy masonry roof. Here the roof is a single element sitting lightly on the steel columns that have been positioned around the circumference to match the support for the glazing. The columns have been deliberately reduced in size and increased in number to minimise the vertical support for the roof, implying a light roof above and allowing as much light as possible through the side walls. The vault is constructed using a single layer of bricks visible on the slab edge to demonstrate its thinness. The roof is finished on the outside with sand/cement render painted white.

As with many of Dieste's structures, emphasis is placed on the continuity of the surface form. The edge slab and cone are connected together seamlessly. In this simple form there is an ingenious precision in the geometry of the structure that eliminates the need for a ring beam and circumferential reinforcing (*see Fig.7.25*). In a conventional dome, the self-weight creates horizontal thrusts that encourage the dome to spread laterally. These thrusts generate tensions around the base of the dome that are resisted either by buttresses, used to transfer the thrust to the foundation or by a strong, continuous ring beam that contains the thrusts at base of the cone,[189] avoiding the transfer of horizontal thrust to the supporting structure below. The ring beam provides a clear delineation to the edge of the dome.

In Parador Ayui, the cantilever of the edge slab eliminates the need for either a buttress or a

Chapter Seven
Typological Variations

ring beam, essential to provide the feeling of lightness that the structure has. To understand how the roof works it is useful to consider the roof as a series of tapered segments inclining in on each other. Without the ring beam the roof segments will tend to fall inwards and flatten out. The upper compression ring provides the horizontal reaction resisting this action, the segments effectively leaning against each other. The ring beam at the base of the cone balances the horizontal reactions from the compression ring and prevents the segment from moving outwards. In Parador Ayui Dieste lets the structure work for itself and uses the overhanging slab to eliminate the horizontal thrust of the cone. The overhanging slab counter-balances the inward inclination of each segment as it tries to rotate about the vertical support. If the weight and span of the overhang were increased sufficiently then intuitively the compression forces in the upper ring would be eliminated and tension would develop at the top of the cone. In other words, the segments would rotate in the opposite direction and the cone would open at the top. By careful proportioning of the dimensions the cantilever of the edge slab balances the weight of the segment. The reactions at the top of the segment, which hold it in place are removed and, hence, the force in the compression ring is eliminated. In accordance with Newton's Third Law,[190] by removing the force in the compression ring the force in the tension ring is also removed and there is, therefore, no need for a ring beam

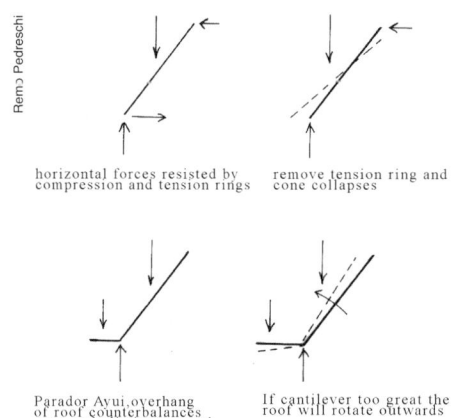

Fig. 7.25. Parador Ayui - structural behaviour.

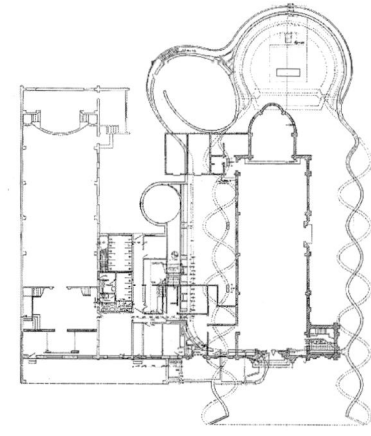

Fig. 7.26. Our Lady of Lourdes - proposed plan and existing church.

and continuity of the surface and lightness of form is assured. Light reinforcement is incorporated within the brickwork to allow for uneven loads.

Nuestra Señora de Lourdes[191]

The project for a new church in the Malvin district of Montevideo was started in 1967 and was never completed. It was certainly the most ambitious of his projects up to that point. The project involved the eventual demolition of the existing church after the phased construction of a new parish church, parish house and parish centre. The new church would be built around the existing building allowing religious services to continue as long as possible during construction. The proposed plan shows the nave of the new church enveloping the old (Figs 7.26 and 7.27). The curved nature of the nave walls illustrates a technical and conceptual progression from Atlantida, in which Dieste explored the potential of the curved wall surface. The plan of the new church has an organic nature as it grows around the old building with the altar situated in a tall gently tapering semi-circular tower around the presbytery (Fig.7.28). The curvature of the wall is more complex than Atlantida, starting from a curved baseline, continuously changing as the walls rise to meet the roof. A modified version of the design was completed 30 years later in Spain for the church of San Juan de Avila in Alcalá de Henares and is described in more detail in the next chapter. The only parts that were actually completed were the parish house and ancillary

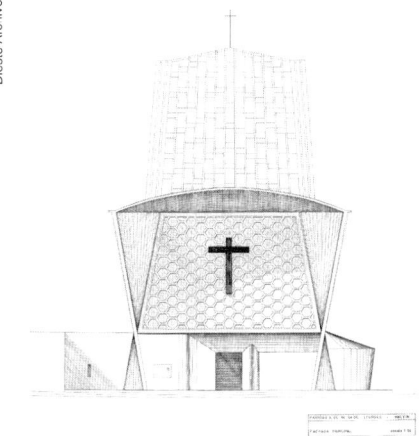

Fig 7.27 Our Lady of Lourdes – proposed entrance.

Fig. 7.28. Our Lady of Lourdes – model.

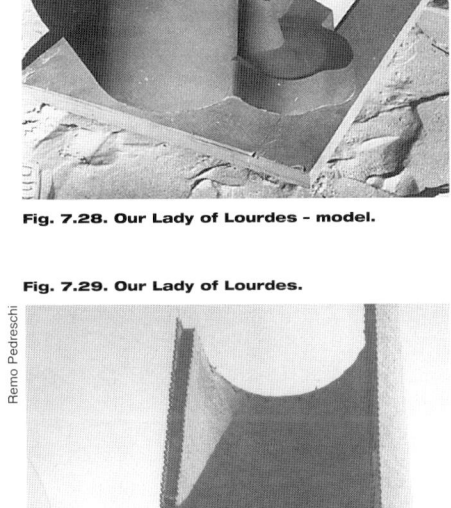

Fig. 7.30. Club Remeros – Salto.

Fig. 7.29. Our Lady of Lourdes.

Fig. 7.31. ADF Wool warehouse – offices.

buildings and the part construction of the presbytery.

The parish house shares some common features with Dieste's own house in the use of the barrel vault and the screened terrace at first floor presents a protective façade to the street. The partially constructed presbytery, with its 'teeth' of un-finished brickwork stands over the old church threatening to devour it should the call ever come (*Fig. 7.29*).

There are a number of other buildings, designed by Dieste, that have not been included in this chapter but continue the theme of typological variation. These include Club Remeros in Salto, further exploring the curved vertical surface and the carefully detailed façade of the office building at the ADF Warehouse (*Figs 7.30 and 7.31*).

Chapter Eight
Crossing the Atlantic

Chapter Eight
Crossing the Atlantic

'*I believe that we must contemplate each problem independently, keeping in mind the conditions and our circumstances and environments, which are so different from the conditions of the developed countries.*'[192]

In his essay, *Technology and underdevelopment*, Dieste describes a common attitude in the developing countries; that, when faced with problems, it is taken for granted that either the solution or the route to the solution is to be found in the developed world. There is an oblivious acceptance of the technology of the developed world, often misunderstanding the context in which it was created or ignoring the principles upon which it was founded. Dieste has purposefully avoided the import of first world technology, in part for economic and practical reasons and in part for reasons of contextual appropriateness. The reinforced brick technology developed by Dieste contrived to suit the needs of Uruguay, using the indigenous skills and materials of the country. That, in combination with his awareness of structure and form, has engendered the unique character and language of his work. His language of construction is open to oxymoronic interpretation where the complex and minimal structural surface forms appear in dialectical opposition to the 'crude and traditional' brick. From the point of view of the developed world, while the aesthetic form of his buildings is appealing, the use of brick may signify a third world technology, an idiosyncratic creation of a maverick designer, with plentiful access to cheap labour that has no place in the developed world. Has Dieste created a technology appropriate only to South America? This chapter considers a series of projects that Dieste, late in his career, has undertaken outside of South America.

A fortuitous sequence of events in the early 1990s saw a reversal of the conventional flow of ideas from developed to developing countries. During this period his work was becoming known in Europe, promoted by exhibitions and invitations to conferences and symposia that tended, rather unsurprisingly, to celebrate the architecture of his buildings rather than their technical merits. The event to have the most significance was the Fifth Congress on the City of Knowledge held in 1993 at the University of Alcalá in Spain. During the congress Dieste was invited to chair a session on the restoration of the Castle of Molina de Aragón. An important theme of the session was the renovation of culturally important sites experiencing rapid economic growth with limited resources to support sensitive reconstruction. His ideas on an architecture that was culturally appropriate and that could be constructed in an economic and technically rational manner were received enthusiastically. As a result his techniques were exported to Spain in a number of projects in and around the city of Alcalá and also inspired research and development into contemporary brick vaults in Germany.

'*The use of brick allowed, ….. , the recovery of a system of construction - even more a general typology of building - which endured in the Alcarrian scene, with its clear Toledan influence.*'[193]

He suggested that brickwork had an appropriate texture for urban renewal, suitable for building contemporary architecture. His ideas had resonance with a theme current in Spanish architecture at that time. Internationally known Spanish architects such as Rafael Moneo, Mariano Bayón and others used brickwork in a variety of large institutional, commercial and residential projects. Differing attitudes to the expression of brickwork have evolved. Enrique Sobejano, in an edition of *Daidalos*,[194] considered the use of brickwork in recent Spanish architecture. He identified two contrasting approaches, which were used to compare directly the work of Moneo and Bayón.

'*Realism and abstraction, heaviness and lightness, are contradictory concepts of a much broader-based dialectic of brick architecture. There are two contradictory currents in the use of this building material. One tends to testify to its load-bearing qualities and the function of the corporeal, durable, tangible and real wall. The other conveys to us the impression of a skin, or a tender surface which no longer subjects itself to the laws of gravity.*'

The load-bearing qualities refer to the representation of brickwork in its traditional sense, as a robust, solid material for the vertical transfer of force or laterally through the thrusts in large, sometimes exaggerated, arches designed by architects such as Moneo. Or as skin in the work of Bayón, in Poligno de Palomares, a housing estate in Madrid. These different interpretations of the material character of brickwork were also suggested by Ochshorn.[195] Dieste's work, by extending the application of brickwork into a wider range of building types, seems to satisfy both conditions, being expressively load-bearing and also lightweight, at times appearing to defy gravity.

Following the symposium, Dieste established a working relationship with the architects Carlos Clemente[196] and Juan de Dios de la Hoz, based in Alcalá, who were keen to study his techniques further. Initially, some small additions were made into projects already underway in Alcalá. The projects were successful and led to the much more ambitious Camino de los Estudiantes, discussed later in this chapter. These events coincided with significant demographic changes happening in the area around Alcalá.

The city of Alcalá is located on the River Henares, 30 km east of Madrid. The city has played an important part in the cultural and social development of Spain. The present city evolved from the Roman town of Complutum and was later renamed, by the Moors, al-Kal'a Nahr, from which the present name is derived. In 1998, it was inscribed by UNESCO as a centre of world heritage, primarily for its contribution to the 'intellectual development of humankind'. It is considered to be the first city developed and built as the seat of a university[197] and played a major part in the development of the Spanish language. It is also considered to be the first example of the ideal city, Civitas Dei (the City of God), providing ideas used by Spanish missionaries in the colonisation of the Americas and other countries. It is also the birthplace of Miguel de Cervantes Saavedra,[198] creator of the most famous character in Spanish literature, Don Quixote de la Mancha, whose statue dominates the main square of the city. The area around the city and following the river in a south west direction is known as El Corredor de Henares. In recent years, there has been substantial economic expansion along the river with a rapidly growing population moving in to take up opportunities provided by the new industries and companies. Many of the existing towns and villages of the region have expanded rapidly and new communities have formed.

In 1991, a new diocese of Alcalá de Henares[199] was created to serve the spiritual needs of the growing population.[200] The new bishop was faced with considerable problems; communities were expanding haphazardly all across the diocese, many existing parishes were no longer the focus of the new settlements and were ill-suited to meet the needs of the laity. Many of these embryonic communities lacked the facilities[201] that bring people together. At least 20 new parish churches were needed in the diocese. The churches would serve both a religious and a pastoral role and include both community and religious facilities. In addition the intention was

Chapter Eight
Crossing the Atlantic

that they be special and yet familiar places, the embodiment of the Liturgical Movement that embraced church design earlier in the century. As always economics were important.

During Dieste's second visit to Spain to work on the Camino de los Estudiantes his collaborators, Clemente[202] and de Dios de la Hoz who also acted for the Diocese of Alcalá, no doubt encouraged by their earlier experiments, were keen to explore the potential of Dieste's techniques to assist with the problem of the diocese's building programme. Dieste (now over 75) responded enthusiastically and, in addition to his direct input, offered the designs for his three churches in Uruguay as a starting point for future projects. The following year, a study trip to Uruguay was organised consisting of a group of architects and builders from Spain who returned convinced that they had found a system of building that had all the qualities they were looking for. At first glance, the idea of using 'system building'[203] in the context of churches would be an anathema to most designers. Can spirituality exist in the repetitive and prefabricated?[204] However, in Dieste's work the diocese saw the potential to develop a construction system appropriate to church building. As described in the previous chapter, Dieste's techniques are both generic and particular - capable of diversity in expression. Two models already existed. The churches at Durazno and Atlantida came from the same genetic pool of construction techniques and shared many characteristics and considerable diversity in others. They share a common approach to the planning of internal space, of the nave and altar being unified, of economics and construction processes but their manifestation and expression are very different.

Three churches were planned based on the designs for Atlantida and Durazno and for the design for the unfinished church at Malvin in Montevideo. Work started almost immediately to adapt the designs of Atlantida and Durazno. In addition to the churches themselves, all three projects were to incorporate additional accommodation to meet the needs for the community, rooms for meetings and classes and offices for the clergy. Dieste y Montañez were involved throughout the process and developed the modified designs, providing the structural engineering calculations and construction procedures for each of the projects.

The parish church of San Juan de Avila, Alcalá de Henares

The church is located on the northern edge of the city surrounded by rising apartment blocks (*Fig. 8.1*). Of the three projects, this is perhaps the most interesting as it is based on the design for the Church of Our Lady of Lourdes in the Malvin district of Montevideo that was never completed. It is also the most ambitious technically, combining the curved walls of the nave, similar to Atlantida, with a tall conical tower forming the presbytery around the altar as used at Durazno. The geometry of curved nave walls is more complex than in earlier projects; the surface of the walls curves and sweeps from side to side and then meets with the inclined wall of the presbytery at the rear of the church. The plan of the church, with the curvature of the walls at different levels, indicates the complexity of the construction (Figs 8.2 and 8.3). Unlike Atlantida, where the walls started from a straight line at the foundation, these walls start from a curved baseline, and then incline either inwards or outwards as the wall rises, forming a series of undulations. As the wall rises the depth of the undulations reduces until approximately one third of its overall height. At this point a horizontal section through the wall reveals a straight line. As the wall rises past this section the undulations of the surface of the wall start to increase, in the opposite direction, reaching a maximum at the top of the wall. The amplitude of the waves of the walls at the eaves is twice the amplitude at its base.

Internally the wall creates a similar feeling to Atlantida, a warm, secure blanket thrown round the congregation. The taller walls of San Juan create a greater feeling of depth to the internal plan, starting from the entrance and merging with the presbytery. The tower of the presbytery acts as a reflector for light from the large cross-shaped window above the nave (*Fig. 8.4*). The church is considerably taller than the church at Atlantida, 11·1 m to the eaves and 23 m to the top to the presbytery tower (*Fig. 8.5*).

The roof over the nave is a Gaussian vault similar to Atlantida and meets the wall in a similar arrangement of curved edge beams. The desire to bring the clergy and the laity closer by bringing together the altar and nave is an important consideration in both churches and different approaches are apparent. In Atlantida, the walls of the nave terminated against a flat gable wall behind the altar. A second, horseshoe shaped wall, was wrapped around the altar to meet the nave and draw both parts together. In San Juan's plan, the nave walls and the connection with the presbytery achieve the same effect in a more decisive manner, both the walls and the ceiling of the nave running into the presbytery which itself curves around the altar. The front elevation (Fig.8.6) is similarly slotted into the nave walls with a band of translucent alabaster filling the gap around.

A large circular stained glass window, 6 m in diameter sits above the entrance (Fig. 8.7). The window was designed by the stained glass artist Carlos Muñoz de Pablos, President of the Association of the Royal Glass Workshop of La Granja in Segovia. The window consists of two layers of glass, one layer convex on the exterior and the other concave on the interior forming a lens shaped cross-section that improves resistance against wind loads. The window is framed in steel, restrained within four concentric circles of brickwork, that in turn sits inside a rectangle of reinforced concrete beams. The beams are 260 mm^2 cross-section and stabilise the elevation against wind loads. Behind the window, above the entrance to the church, is the choir loft. The stairs up to the choir loft are constructed in reinforced brickwork. The balustrades to the stairs are reinforced solid brick panels inclined to the angle of the stairs themselves. This is an improvement in the detail used at Atlantida where the balustrades were made from discrete ribs of bricks and have been found to be too flimsy.[205]

The walls of the nave are constructed using two layers of brick and incorporate a layer of polyurethane insulation like the walls at Montevideo Shopping. The roof also incorporates a layer of insulation protected with a layer of sand and cement render.

In a similar manner to Atlantida, the

Fig. 8.1. San Juan de Avila.

Chapter Eight
Crossing the Atlantic

apparently perverse geometry of the cross-section has an underlying structural function and can be seen as a natural progression from Atlantida as the height of the walls increase. The walls of San Juan are considerably taller than those at Atlantida although the span of the roof is almost identical. Therefore, the walls of San Juan have to carry much greater bending moments. In chapter 4, the cross-section of Atlantida was compared to a portal frame with a pinned base,[206] its shape matching the pattern of bending moments. The structural form of the cross-section of the nave in San Juan also follows the pattern of bending moments. In this case the analogy is for a portal frame with a fixed base. If a portal frame is designed such that the base is fixed,[207] i.e. it is capable of transferring bending moments to the foundations, the bending moments in the wall reduce in comparison with a wall where the base is 'pinned'. As the section of the nave deforms under loading and one envisages the changes in geometry of the wall[208] there is a change in the direction of curvature of the vertical elements (Fig. 8.8). The change in curvature occurs at the point of contra-flexure.[209] The position of the point of contra-flexure depends on the relative height to span of the frame and the relative thickness of the vertical and horizontal elements[210] but is typically located at approximately one third of the height from the base. The change in curvature occurs as bending moments are transferred to the foundation and in turn help to reduce the moments in other parts of the cross-section. Thus, the section created by the undulations of the wall and roof of San Juan de Avila follows the bending moment diagram (Fig. 8.9). The diagrams, drawn to the same scale, also compare the bending moments for both churches. The bending moment diagram under vertical loading, above the point of contra-flexure for San Juan, is almost identical to Atlantida. It appears as if the section of Atlantida has been raised onto buttresses,

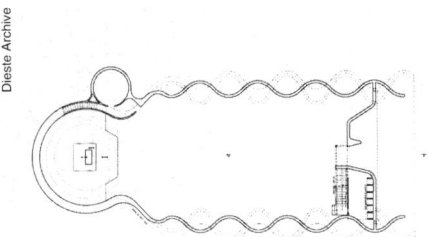

Fig. 8.2. San Juan de Avila - plan.

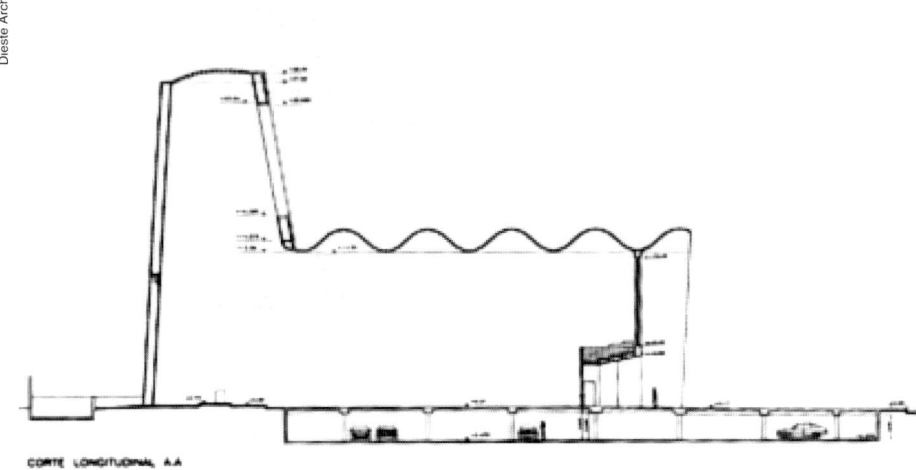

Fig. 8.3. San Juan de Avila - section.

Fig. 8.4. San Juan de Avila – from altar to rear of church.

created by the undulations below the point of contra-flexure. When wind load is applied to the nave walls the effect of the changing geometry of San Juan becomes even more significant. By comparing the section of the church of San Juan with a wall of the same height, having the same section as the church of Atlantida, the maximum bending moment along the height of the wall is reduced more than three times by adopting the former condition.

The walls of San Juan de Avila, although proposed some 30 years earlier, are even today innovative. How to build tall brick walls in single storey buildings has been discussed in

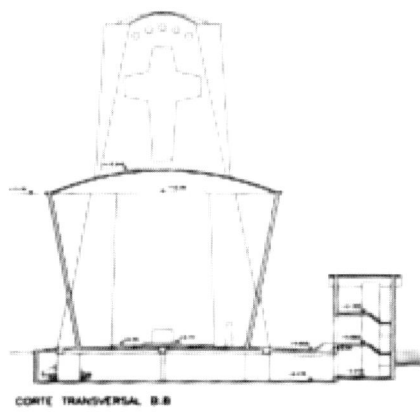

CORTE TRANSVERSAL B.B

Chapter Eight
Crossing the Atlantic

Fig. 8.5. San Juan de Avila - presbytery.

Fig. 8.6. San Juan de Avila - front elevation.

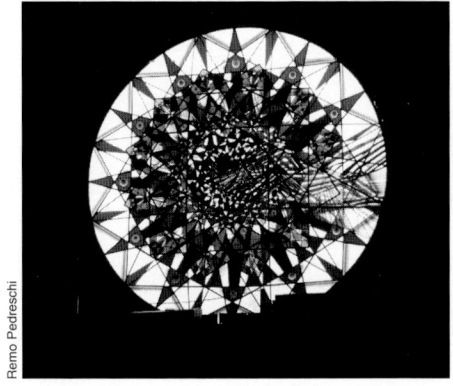

Fig. 8.7. San Juan de Avila - stained glass window.

previous chapters and there are several ways to do it. The alternatives, such as the fin or diaphragm wall, tend to assume that the wall is, in structural terms, independent from the roof, and is designed to resist the forces on its own, either as a vertical cantilever or a propped cantilever. In either case, the stresses for comparable heights would be much greater than Dieste's constructions that see the roof and walls as integral parts of the structural section. To deal with these stresses, the wall itself would have to be either thicker, using more materials, or the amplitude of the curves would have to increase. The conventional solution is to use walls with vertical surfaces and constant horizontal cross-section. The cross-sectional dimensions are based on the structural requirements needed to deal with the maximum bending moments at their base — wasteful in material, less expressive and reducing the effective floor area of the building. Dieste's design is innovative therefore on a number of counts:

- the continuity of structure between wall and roof, allowing them to act together structurally
- the manipulation of the wall surface, shaped to follow the bending moments
- the rigid connection at the base of the wall, allowing reduction in bending moments.

Curtin,[211] in considering his work on diaphragm walls, made an analogy with the simple rectangular cross-section of conventional walls and the improved structural cross-section of the diaphragm wall[212] with that of a rectangular beam and the structural efficiency of an I-section beam. Dieste has taken at least two steps further forward in the use of structural brickwork by manipulating the surface of the wall itself. By developing continuity between wall and roof he avoids the need for the 'improved section' creating a purer structure. In truth, this is a remarkable manipulation of the surface geometry to control structural behaviour, one of the most sophisticated and confident examples of the use of structural surface form. The steps taken by Dieste are exponential rather than incremental leaps forward in concept and realisation. If the Gaussian vault is the ideal structural surface form for a roof, then Dieste's curved surface of San Juan present the ideal

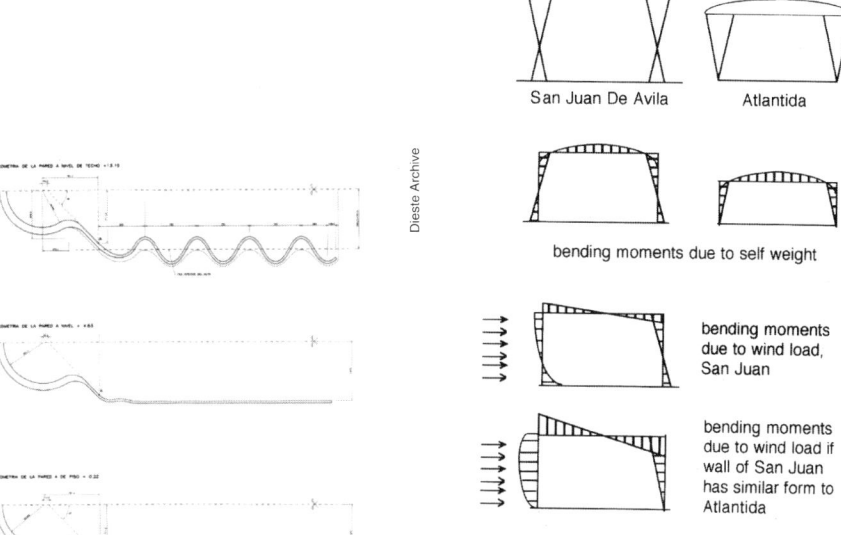

Fig. 8.9. San Juan de Avila - bending moments.

Fig. 8.10. San Juan de Avila - under construction.

Fig. 8.8. San Juan de Avila - geometry of wall.

surface for the a tall wall.

The geometry of the walls of San Juan was determined in the same way as Atlantida, using straight lines, set out from a curved baseline and then displaced to follow the curved plan at the eaves. The side walls of the nave were constructed first and then followed the walls of the presbytery (Fig.8.10). The junction between the nave and the presbytery requires the careful resolution of two different forms; an intricate manoeuvre that brings together the cross-sections of the nave and the presbytery in an apparently seamless manner (Fig. 8.11). Internally, there is no hint of this junction, with the brick bonding on horizontal courses continuing between both parts (Fig. 8.12). The eye trained in construction

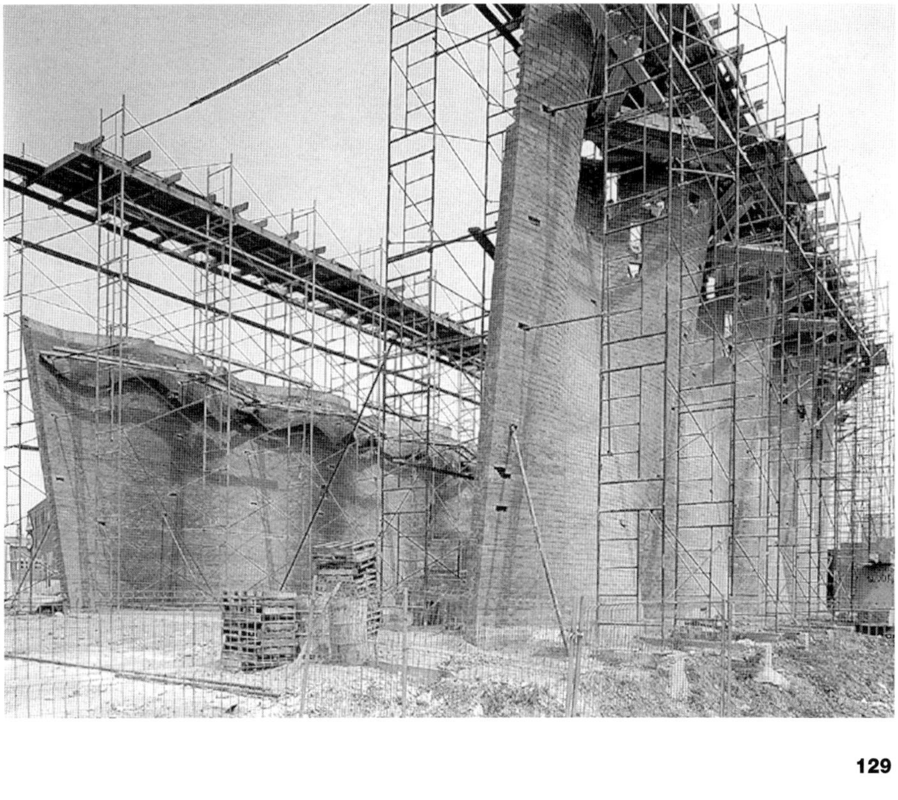

Chapter Eight
Crossing the Atlantic

Fig. 8.11. San Juan de Avila – junction between nave and presbytery.

looks for the seam but does not see it and only then understands the craft used in its making. The two curved surfaces form a common inclined edge. The walls meet at a point where the nave wall inclines inward. This line, which can be traced on the front elevation continues upwards past the level of the nave roof and forms the edge of the front elevation to the presbytery that is punctured by the large crucifix window. The window is 7·5 m in height and sits in a 900 mm wide diaphragm wall. The weight of the window and wall is carried by a composite box shaped beam consisting of reinforced brick webs and concrete flanges, which form a 2 m deep construction that spans across the nave. The beam is not apparent internally as the inner skin of brickwork continues below the bottom

Fig. 8.12. San Juan de Avila - interior.

flange of the beam and curves to meet the soffit of the nave of the roof.

Two vertical concrete beams, on either side of the cross, run from the top of the box beam to the concrete slab at the head of the cross, which in turn, spans onto the side walls. The external surface is rendered and painted.

The wall of the presbytery curves round the altar, inclined inwards. The wall is constructed as a diaphragm with two skins, separated by brick ribs, similar to San Pedro in Durazno. The roof over the presbytery is a shallow dome-shaped concrete slab, perforated with 126 circular windows formed using large clay pots. A circular baptistery is located on the south elevation between the presbytery and the nave. The ancillary accommodation is housed in a separate building parallel to the south side of the church. The building has a linear plan, intended to rise three storeys, though at present completed only to the ground floor. An analysis of the construction costs for of the church was published in *Costes + Datos de Edificación*.[213] The total cost of the church, including the community building and excluding fees and taxes was 588 000 Euros.[214] The cost of the reinforced brick walls and roof of the church were 451 Euros per square metre of church.

Nuestra Madre del Rosario[215]

The first of Dieste's designs to be constructed in Spain was the parish church of Nuestra Madre del Rosario in the town of Mejorado del Campo, located along the Henares Corredor, south west of Alcalá (*Fig. 8.13*). The problems of the town are clear, as it sprawls outwards towards the flat countryside, with a mixture of light industry and residential areas, organised in a rectangular grid. The town feels as if it has outgrown the centre of the original village, which now cannot keep up with the new suburban housing estates sprouting around its periphery. The visitor, arriving by public transport for the first time, may find it difficult to orientate himself as the town lacks an obvious centre, either commercial or civic. The expansion of the town seems predicated on industrial and residential growth, without regard to civic values and amenities that should accompany such growth. Many of the inhabitants have moved there recently.[216]

The church, Nuestra Madre del Rosaria, is situated near the centre of the town facing onto low-rise apartment buildings. The design is based on the church of San Pedro in Durazno, though with considerable modification. The new church is approximately one third smaller in size. The context of the two churches is quite different. San Pedro being reconstructed from the remnants of an older church. The scale and dimensions of San Pedro were taken from the footprint of the original church and the existing façade was retained. San Pedro was surrounded on all sides by existing buildings and, hence, Dieste's design was primarily intended to be experienced from the inside. The Spanish church, however, located on an open site, does

Chapter Eight
Crossing the Atlantic

not have the constraints that helped form San Pedro so specifically (*Fig. 8.14*).

In the new design the entrance to the church is intended to be through the community building that forms a long façade to the street with its axis perpendicular to the church. This building serves a function similar to the façade of the original church of San Pedro, providing a mask to the church from the street. At the time of writing the community building had not been constructed and the church consists at present of the nave and presbytery entered directly under the rosette window, which forms a temporary elevation to the street. Once the amenity building is complete, the rosette window will be lit through windows in its façade and roof.

On entering the building there is an experience, similar to San Pedro, of controlled light, although the overall level of light is greater with less contrast between light and shade, particularly between the nave and the side aisles (*Fig. 8.15a & b*). This appears to be due in part to the rosette window occupying a proportionately larger area of the front façade and the bricks themselves, which are lighter in colour than the darker, more reddish Uruguayan bricks. Once the front façade is completed the light source to the rosette will be diffused through the roof lights and windows on the façade of the community building and reduce the overall level of light to the nave. The altar and the presbytery are almost identical to the Uruguayan church with light reflected from the sides of the tower onto a suspended crucifix.

Fig. 8.13. Nuestra Madre del Rosario.

Fig. 8.14. Nuestra Madre del Rosario.

Two small rooms, one forming the sacristy, are located behind each side aisle and accessed directly from the aisle.

In the construction, there are a number subtle but important differences between the two churches that highlight the uncertainty of the Spanish builders compared with the confidence shown by Vittorio Vergalito and his workers at Durazno. In the original church considerable efforts were taken to ensure continuity of the internal surfaces by controlling the bonding of the brickwork. Where the inclined wall of the nave meets the sloping ceiling of the aisle roof a discontinuity in the laying of the bricks, not apparent in Durazno, is clearly noticeable. A more significant difference is at the junction between the end wall of the side aisle and the inclined wall of

the presbytery. The bricks are laid in a vertical stack bonded pattern, with vertical joints between stacks. At the junction between the walls the vertical stack bond has to reconcile with the inward inclination of the presbytery and nave walls. The last stack consists of carefully cut tapered bricks that, rather unfortunately, breaks the rhythm of the stacks. In Durazno, the stack bonding is inclined at the same angle as the nave wall, avoiding cut bricks and maintaining the continuity of the surface - clearly more difficult to build, but a detail which perhaps places flawlessness over pragmatism. The roof of the nave at Nuestra Madre del Rosario is supported from a reinforced concrete beam expressed externally, spanning the full length of the nave, from the entrance to the church to the tower of the presbytery. This is a considerable simplification

Fig. 8.15a & b. Nuestra Madre del Rosario - interior.

Chapter Eight
Crossing the Atlantic

compared with the prestressed brick wall in Durazno. The detail of the nave wall and its complex system of prestressing in Uruguay was probably too great a challenge for the first church project in Spain and a more familiar technique was used.

A new detail in the church is the incorporation of the Stations of the Cross[217] into the walls of the side aisle *(Fig. 8.15b)*. These walls are double skins of brick, the external skin being vertical and the internal skin inclined at the same angle as the nave walls. The Stations of the Cross are represented by 14 square windows created by the insertion of sheets of translucent alabaster built into each skin. Fluorescent lighting within the wall is used to illuminate the stone. The cavity is insulated and contains the down pipes for the roof drainage system.

In some respects Durazno, with its planar geometry, was simpler to construct than the curving walls of Atlantida, although some of the reinforcing and prestressing arrangements were more difficult. By modifying some construction details to eliminate unconventional prestressing and incorporating more familiar techniques of reinforced concrete for some critical elements the Spanish builders started with a comparatively straightforward design. Such a compromise is inevitable. The good teacher always knows more than he can teach. Years of experience and confidence gained during Dieste's long career cannot be replaced by even the most careful study of construction techniques and design.

Fig. 8.16. La Sagrada Familia.

La Sagrada Familia, Torrejon de Ardoz

The second church to be constructed was the Church of the Holy Family in Torrejon de Ardoz, a town on the Rio Henares mid-way between Alcalá and Merjorada del Campo. The town has many of the characteristics of Merjorado, undergoing considerable expansion, old buildings being refurbished, new buildings under construction and a population swelling with incomers. The church (*Fig. 8.16*) is located in a new district of the town bounded on one side by a new sports stadium. On the other side are open sites awaiting development and in front, across the street from the church, are new apartment buildings. The design of the church was adapted from the church at Atlantida, the best known of Dieste's works. It is a pretty faithful reproduction with only a few variations. The baptistery is located inside the church and followed the prevailing guidance from the Liturgical Movement for the ceremony of baptism to be a celebration for and witnessed by the whole congregation. The campanile has been retained; but it is now positioned in front of the church to one side. It is also used to conceal a steel flue. The ancillary accommodation, not yet built, is to be placed along one side of the church in a separate building, three storeys in height. Internally, the church is very similar with a side chapel and sacristy in the residual space between the curved wall defining the presbytery and the side walls. The church is heavily adorned with large pictures of stations of the cross and an elaborate crucifix over the altar (*Fig. 8.17*). In comparison with the church at Atlantida, these seem to conflict with the building itself, disrupting the internal surfaces. However, they are also evidence of a congregation that is alive and values the church. There are some differences in construction between the two churches. In common with the other Spanish projects a layer of insulation is now incorporated into the walls and roof. The roof vault is covered with a second skin of bricks rather than rendered in the original.

Camino de Los Estudiantes

The first major project involving Dieste in Spain was the Student Promenade[218] at the campus of the University of Alcalá. The campus sits on the edge of the city and is reached by its own rail station. The Camino de los Estudinates is a grand, romantic gesture, appropriately Quixotic in its conception. This very long covered walkway, 1·5 km overall, consists of a series of 52 double cantilever barrel vaults, each one 30 m long, overlapping and twisting up a gentle slope toward the University. Interspersed along the route are three very large conical brick structures forming major insertions into an otherwise monotonous landscape (*see Fig. 8.18*). They have no other function than the shelter of students on their way to their studies. The vaults are supported at their centre on a pair of reinforced concrete columns, cantilevering 15 m on either side. The height of the columns vary from vault to vault, allowing them to overlap and rotate on plan to form a

Chapter Eight
Crossing the Atlantic

route that meanders gently towards the University (*Figs 8.19 and 8.20*).

The design of the vaults follows the principles of similar free-standing barrel vaults in Uruguay. The cross-section of each vault is a catenary 1·6 m high by 4·2 m wide, the lateral thrust is resisted by tapered edge beams. The vault is prestressed longitudinally along its crown using nine, 16 mm diameter cables with a total prestress force of 17·8 tonnes. The bricks have a minimum compressive strength of 20 N/mm^2, the maximum working stress under snow loading is 5 N/mm^2.[219] The vault is 8 cm thick, increasing to 10 cm for a distance of 5 m on either side of the columns to incorporate prestressed cables looped diagonally over the vault that resist the high shear forces around the column.

The vaults are constructed in two parts using a 15 m long prefabricated form on a steel framework that sits on rails. Once the reinforced concrete columns have been completed, the formwork is moved into position, the bricks and reinforcement are placed and then grouted (*Fig. 8.21*). The formwork is struck when the mortar (a one to three cement and sand mix) reached a compressive strength of 5 N/mm². Normally, this strength would be reached one day after filling the joints. The weight of the half vault is taken by props running along the edge beam of the roof. Inclined props are used to support the horizontal thrust from the edge beam. The formwork is then rolled along the tracks to the other side of the columns and the second half of the vault is

Fig. 8.17. La Sagrada Familia - interior.

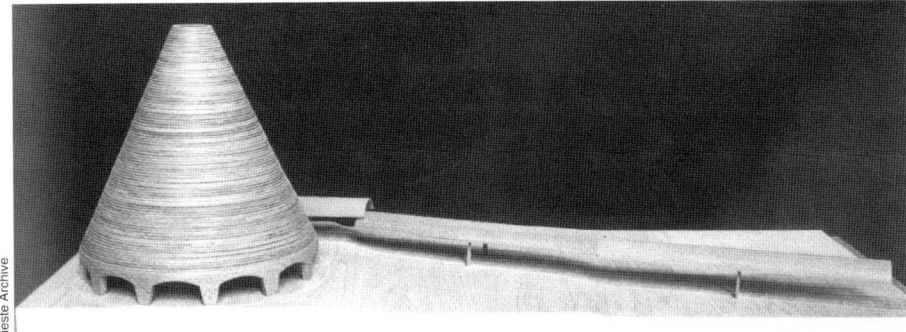

Fig. 8.18. Camino de los Estudiantes - model.

constructed and propped in the same way (*Fig. 8.22*). Once both ends have cured, the vault is prestressed and a protective layer of sand–cement render is applied and cured and, finally, the temporary props are struck. In the notes for the design, Dieste y Montañez suggested a construction programme on a repetitive cycle of five weeks. This assumes that the foundations and columns are completed in advance.

1st week   Erect formwork for half of vault one and lay bricks and strike formwork.

2nd week   Erect formwork for second half of vault one and lay bricks. Apply finish render to the first half of vault one.

3rd week   Strike formwork for second half of vault one and erect for first part of vault two and lay bricks. Apply finish render to second half of vault one.

4th week   Strip formwork and erect for second part of second vault and lay bricks. Apply finish render to first half of vault two.

5th week   Prestress vault one. Prestressing would only take place once the mortar had attained a strength of 20 N/mm^2. Apply render to second half of vault two.

The construction cycle is predicated on maximising the use of the formwork, once the cycle had started one vault would be completed every three weeks using one set of forms.

Chapter Eight
Crossing the Atlantic

Fig. 8.19. Camino de los Estudiantes - covered walkway.

Fig. 8.20. Camino de los Estudiantes - overlapping vault.

Fig. 8.21. Camino de los Estudiates - formwork.

The two conical brick structures are referred to as arbours and are intended to be shady retreats[220] for rest and relaxation, places for discussion and contemplation. These large truncated cones are over 20 m high and 25 m in diameter, similar to Victorian glass blower's cones. At present, one can only imagine the nature of such a large enclosed space. The cone rests on a circular beam supported by reinforced brick piers. The beam is reinforced circumferentially to resist lateral pressure from the cone. The beam itself is supported by shallow arches spanning 4·6 m onto the piers. The cone inclines inwards at 60° to the horizontal. The bricks are laid in horizontal courses that corbel inwards and have two layers of reinforcement running between the bricks parallel to the slope of the cone. Circular steel reinforcement is placed in every fourth course.

A similar structure was designed recently by the English architect Tim Ronalds for the Landmark Entertainment Centre, Ilfracombe.[221] The sizes of the cones are similar, 23 m in height and 20 m in diameter, but the construction was quite different. Ronalds had to provide an environmentally conditioned and weathertight envelope for the auditorium inside. The walls are constructed in inclined brickwork, rather than the corbelled construction of Dieste, and incorporate an insulated and waterproof layer to the inner leaf of the wall, protected by an outer leaf of white brickwork.

At the time of writing, work has stopped on the project in Alcalá due to some 'technical misunderstandings' with the local Spanish engineers, however, it is planned to complete the project. These problems certainly did not discourage the use of Dieste's methods in the three subsequent churches which have been completed. The construction of San Juan de Avila is of particular importance, as it demonstrates an important progression in the development of Dieste's ideas of construction and church design, developing further the construction of curved walls. It also marks a rapid development in the construction skills of the Spanish builders. As a method for the economic construction of churches the case seems proven. According to Carlos Clemente,[222] the three churches were between 15 and 25% cheaper than conventionally constructed churches of similar size and quality

Fig. 8.22. Camino de los Estudiantes - temporary props to vault.

using steel or reinforced concrete. The church projects were awarded the Premio Iberoamericano by the Spanish Ministry of Public Works. The Diocese of Alcalá has since completed two more churches, Santa Cruz in Coslada on the Henares corridor and Nuestra Senora de Belén in Alcalá. These are new designs by Clemente and de Dios de la Hoz[223] using the same construction methods, and calculated by Dieste y Montañez.

Insitut für Tragwerksenturf und Bauweisenforschung[224]

An interesting footnote to this chapter is the recent development of a Gaussian vaulted roof system recently developed by the Insitut für Tragwerksenturf und Bauweisenforschung at the University of Hanover in Germany (Fig. 8.23).

The designers explicitly acknowledge the work of Dieste as being the inspiration for their ideas. Dieste has a long association with Germany; his wife, Elisa, is from Germany and he speaks German and, moreover, the first exhibition of Dieste's work in Europe was in Berlin at the Hochschule in 1991. This was also the occasion of Dieste's address to the VIth International Conference on Brick Masonry. The system adopts the form of the Gaussian vault but modifies the construction process to suit German practices. The purpose of the changes are to eliminate the use of bed joint reinforcement, which, according to German standards, has insufficient corrosion protection and to reduce the complexity of the formwork.

The vaults are assembled from prefabricated elements. Each prefabricated segment is built vertically like a conventional wall using stretcher bond against a series of plywood patterns, cut from flat sheet, thus describing the curvature of the vault using a series of simple two dimensional planes. Once the segments have reached the required strength, they are placed on the formwork using purpose-designed lifting equipment. A typical segment weighs 450 kilos. The formwork for the vault is simpler than Dieste's as it supports large hardened components rather than individual bricks and it only has to prop rather than describe the complete surface (Fig. 8.24). A 25 mm gap is left between each segment and is grouted once the vault is complete. These construction methods replace some of the more labour intensive activities of Dieste's process such as fixing the steel and the construction of a full formwork for the vault, although there is the need for more mechanical handling. The units would normally be prefabricated off-site, allowing rapid site erection. It is a logical extension to Dieste's argument in favour of brick over concrete when he states that in brick vault construction over 90% of the material is already hardened. In this case over 99% of the material is already hardened. This method of construction could easily be used with precast concrete segments. However, as each segment of the vault varies in cross-section according to its position in the vault, there would need to be a separate form produced for each segment - each form being a complex fabrication describing a continuously varying double-curvature surface. A series of full-scale prototype shells have been constructed and tested. At the time of writing, the system is under patent review.

The contrasting approach of the two countries to Dieste's technology is worth remarking on. In Spain, an empirical approach has been adopted, inspired largely by the architecture that it has been shown capable of providing, its acceptance ratified by meeting other criteria such as cost and practicality. One can sense a warm and emotional response to Dieste and an enthusiasm to explore the potential of his methods in undertaking a series of complex projects, albeit adaptations of existing designs. In Germany, a much cooler approach has been

Chapter Eight
Crossing the Atlantic

Fig. 8.23. German Gaussian vaults.

Fig. 8.24. German Gaussian vault - under construction.

pursued, that recognises directly the elements of the idea behind the Gaussian vault, the combination of structural form, construction process and material to create a structure that consists of 'Only the essential'.[225] They have taken this idea and stripped it, then reformed and adapted it to have economic and practical viability in a highly developed, industrialised economy.

This chapter started by considering the application of Dieste's reinforced ceramic technology in a developed European country used to and comfortable with contemporary construction methods in reinforced concrete and steel. Dieste has demonstrated that reinforced brick is a contemporary structural material, as effective and economic as steel or concrete. The pursuit of this idea and leading to 'the inevitable invention' has taken the work of Dieste across the Atlantic and firmly into the heart of Europe.

Conclusion

This book has endeavoured to illustrate the work of the engineer Eladio Dieste. It has presented only a part of his extensive œuvre. Throughout his long career - over 50 years - Dieste has carried out many diverse projects. These include major works of civil engineering and the design of mechanical equipment and furniture. Certain themes recur throughout his work due to the consistency of his approach.

The exacting standards that his work meets he has set himself. Perhaps this is in part why his work is not so well known; as he has not actually promoted it heavily. He does not present a panacea for the construction industry nor does he promote a new language of architecture, but he has created the means of an effective and appropriate method of building that suits the economic and cultural requirements of his own country. There are many areas of his life, work and philosophy that merit further study.[226] This book has concentrated on the ideas and the technology that underpin his buildings and architecture. Dieste is an engineer and builder who has obtained greater recognition as an architect. Since the Industrial Revolution there has been debate, at times with friction and certainly misunderstanding between architects and engineers on their respective roles and responsibilities.

Ove Arup, in his address to the Royal Institution of British Architects, on receiving the Institution Gold Medal, considered the relationship between the architect and engineer. He asked the question when does an engineer become an architect and vice versa? How do their roles and relationships intertwine? He believed that architecture should be organised from three different viewpoints that he termed the **A–B–C** disciplines: **A** for art, **B** for building technology and **C** for commodity. All buildings should consider the importance of the three elements. He was restating the Vitruvian ideals of Firmness, Commodity and Delight, and dividing the responsibility for each element between the architect and the engineer. His address was clearly related to his own, usually satisfying, but occasionally difficult, experiences. He did not suggest a merging of the disciplines but rather the maintaining of the rigour of both, tempered with greater understanding. He believed that engineers should have a broader awareness of and an enthusiasm for architecture, attitudes that he himself clearly possessed.

Most engineers, who are noted for their contribution to architecture of the twentieth century also agree that art and technology must intertwine. Eduardo Torroja expressed this view in his book, *Philosophy of structures*

'... *to what extent this aesthetic factor can be sacrificed to factors of economy will be a matter for consideration in each individual structure. But even so the effect of aesthetics cannot be considered initially in every case even if it is later decided to disregard it.*'

Where would one place Dieste in this debate? In all his work, there exists a strong architecture and he has frequently been described as an architect[227] although he does not claim such as title for himself. When asked if he considered himself an architect, he replied:

'*No. In order to become an architect one needs an academic training which I do not have. At the least I am a builder who can deal with some very simple architectural problems.*'

Dieste believes that there is a corpus of knowledge that defines the core and fundamentals of a discipline that should be learned with rigour and deep understanding - the way he learned engineering at the University of Montevideo. Dieste is not slavish in his promotion of himself as an 'engineer'. He does not promote his work as exemplars of an architecture that an 'engineer' can produce. Instead, Dieste sees art as an intrinsic part of engineering, whether he is designing buildings or machines.[228] The conventional architect–engineer relationship presents a false opposition to the way that he thinks. He did not design buildings because he was frustrated by such relationships. If anything, it was through frustration with the insensitivity of the sponsors of buildings in ignoring humanity and through his concern for the human spirit in buildings that he was drawn to the art of what could be called 'architecture'. These feelings gained strength during the design of the church at Atlantida. He adopts a moral position towards architecture, believing that careful architecture is essential in a humane society.

'*Architecture is the shaping of space -

something that is profoundly related to human life and deeply affects man's happiness.'

His buildings, therefore, have not been created out of a desire to be an architect, but, that form, light, space and the human scale must be present in buildings irrespective of who designs them. Jean Prouvé once said:

'Architect or engineer, what does it matter. It is the building that counts.'

Lack of formal architectural training may have been an advantage in some ways. Dieste has not been encumbered by the formalities of architecture, the allegiance of styles, being a Modernist or Postmodernist, of taking a formal doctrinaire position on architecture. In one sense, there is an engineering approach to his buildings in that he defines certain criteria at the outset of a project that might conspire to form a 'correct solution'. He asks certain questions. How is it made?, how does it work? what does it do? and what should it be? These questions were apparent in the design for the church at Atlantida. He was asked only, 'How is it to be made?' by the client consciously wishing to avoid the services and no doubt costs of an 'architectural' solution. But Dieste would not reply until the other questions were answered, eventually by himself. In answering these questions, Dieste brings all his experience together - cultural, social, aesthetic and spiritual. His technical skills are the enabling skills that bring this experience together in the design. As an engineer he does not 'play with trains', his inventions are not experiments for their own sake. All of the structural innovations for which he is responsible have been driven by the desire to satisfy broader needs, in what he has termed 'cosmic economy'.

He is concerned with the harmful effects that the pressure to develop the economy of Uruguay, to follow an industrialised western economy may have, especially in its disregard of the lower classes, 'La gente sencilla'. Rather than adopt a Luddite approach, he has manipulated technology to suit the needs of his country, creating a way of building that has familiarity and expressiveness for the people. Dieste has taken the idea of construction using reinforced brick firmly into the modern world and applied the technology to a wide range of buildings in a way that is neither slavish nor obdurate. Each building balances and adjusts each of the three constituents of the Arup's ABC disciplines as needed, whether in a church, silo or warehouse.

Dieste has many similarities with other distinguished engineers of the twentieth century, in particular Freyssinet, Torroja, Candela and Prouvé. The development of construction techniques and his understanding of the potential of reinforced and prestressed brickwork parallels Freyssinet's invention of prestressed concrete. Some of Dieste's prestressing techniques, such as looped wires, could easily be applied effectively to concrete. His exploration of structural form, the development of the free-standing brick vault and the Gaussian vault stand shoulder to shoulder with the new forms of reinforced concrete shells developed by Torroja. Torroja actually used brick vaults early in his career but according to a recent biography[229] he was unable to see great possibilities in them.

'The forms of ceramic vault were greatly limited.'

Torroja was concerned primarily with the development and expression of structural surface forms using the plastic free-form qualities of reinforced concrete. He believed that thin concrete shells were the most honest expression of the material nature of concrete. He tries to make the surface speak itself in a natural manner, rather than the formal expression of the lines of structure used, for example, by Pier Luigi Nervi in the Palace of Labour.[230] With Candela, Dieste shares a similar need to use materials and construction methods that are appropriate to his country and both have taken on the role of builder as a means to the realisation of the structures. Dieste, on Candela, says:[231]

'I think Candela's work is a case of faithfulness to the exploitation of a very rational and logical idea within the Mexican environment.'

On the surface, Jean Prouvé, innovator in metallic buildings and structures, highly influential in the modern architecture and High Tech in particular, may seem to have the least in common with Dieste. They share similar attitudes to manufacture and construction. They also share the same respect for the people who made their buildings and a concern for the potentially damaging effects of increasing industrialisation.

Conclusion

They both believe that the form and beauty of an object are a function of its making and that all projects should start with an understanding of the specific needs of the project. Prouvé's[232] comment is unquestionably appropriate to Dieste:

'The pursuit of this idea, leading to 'the Inevitable invention' that has taken the work of Dieste across the Atlantic and firmly into the heart of Europe.'

Dieste has created a unique and extensive collection of buildings. The impact of his work and ideas is expanding into Europe for both pragmatic and architectural reasons (as if the two could be separated). The work has been driven by honesty to materials and structural form, enmeshed with creative and moral responsibility. This is the underlying truth, the 'cosmic economy' of his work.

Endnotes

Chapter One

1. In Spain and Germany, see chapter eight 'Crossing the Atlantic'.
2. The official name of the country is the Oriental Republic of Uruguay, originally called La Banda Oriental (the east bank). The name underscores the geographic significance of Uruguay between the two countries, East of Argentina, the east bank of the river Plate.
3. Approximately 50% of both imports and exports are with Brazil and Argentina.
4. In 1998 the population of Montevideo was 1·4 million and of Uruguay was 3·15 million. This represents one of the largest proportions of a country's population in its capital of any country.
5. Named after the Uruguayan Hero, Jose Gervasio Artigas, who led the revolt against Spain in 1811.
6. Grampone J. (1996). *Eladio Dieste, Engineering Master.* Unpublished thesis, Montevideo.
7. De Torres C. (1992). El Taller Torres-Garcia's, The School of the South and its legacy. University of Texas Press.
8. See Mergenat J. P. (1994). Arquitectura Art Deco En Montevideo (1925-1950). Mercur S.A.
9. Torrecillas (1997).
10. Grampone J. (1996). *Eladio Dieste, Engineering Master.* Unpublished thesis, Montevideo.
11. One child died in infancy.
12. Castedo (1969), p 284.
13. Bullrich (1969)
14. Grampone J. (1996). Eladio Dieste, Engineering Master. Unpublished thesis, Montevideo.
15. As related by Eduardo Dieste.
16. A term used by Eduardo Dieste to describe their site agents.
17. Eduardo Dieste, letter to Antonio Jimenéz Torrecillas, August 1998.
18. Eduardo Dieste, letter to Antonio Jimenéz Torrecillas, August 1998.
19. Dieste is referring to the emotions expressed by some workers at the completion of a complicated and bold construction. *Arte, Peublo, Tecnocracia* (essay on Art, People, Technocracy by Eladio Dieste).
20. Dieste is a partner in this firm.
21. See chapter eight.
22. Torrecillas (1997). In Spanish and English.
23. Eladio Dieste in an essay *Architecture and Construction,* in Torrecillas (1997).
24. Eladio Dieste in an essay *Architecture and Construction,* in Torrecillas (1997).
25. Eladio Dieste in an essay *Architecture and Construction,* in Torrecillas (1997).
26. Attributed to Louis Khan (1903-1974) in the early 1970s.
27. Ochshorn (1999).
28. There are, of course, exceptions. For example, lightweight fabric structures that rely on the three-dimensional surface form for stability.
29. Dieste (1994a and b).
30. *Technology and underdevelopment and Art, People, Technocracy.*
31. Castedo (1969), p 275.
32. Moloch; the name of a Canaanite idol to which children were sacrificed, hence an object to which sacrifices are made. OED.
33. Sinha and Pedreschi (1992).
34. Bayón and Gasparini (1979). Dieste, in a long interview, explains his attitude to materials.
35. The Brick Institute of America Technical Notes Nos 17, 17A, 17L and 17M.
36. The Church at Atlantida as featured in Architectural Review (1961), pp 173-175 and Architecture D'Aujordhui (1961), pp 88-89.

Chapter Two

37. Dieste uses the word tympanum to describe the infilled space between the intrados of the vault and a horizontal line between valleys. Paduart refers to this as the gable.
38. Referring to vaulted structures, Mainstone (1975).
39. Indeed in the studio of Dieste y Montañez, he is often referred to as the master.
40. Mainstone (1975).
41. See Antony Hunt in Macdonald A. J. (2000).

Endnotes

42. Pozzolanic concrete was used extensively by the Romans, most notably in the large domed vault of the Pantheon. According to Robert Mark in his book *Architectural Technology* (1993) the use of concrete in the Pantheon was not a major structural innovation as unreinforced concrete behaves similarly to masonry but its use was based primarily on economic and constructional factors.

43. For an account of the design of Sydney Opera house see *The art of structural design*, by Alan Holgate (1986).

44. See Mark (1993). Robert Mark gives a detailed account of structures up to the Scientific Revolution.

45. See Mark (1993). Robert Mark gives a detailed account of structures up to the Scientific Revolution.

46. A catenary is the natural shape an arch or cable will take to ensure that the force will be in either direct tension or compression. It is considered the structural form that uses material most efficiently.

47. In properly designed barrel vaults the deformations will be small.

48. Schodek (1980), p 423.

49. Salvadori and Norton (1990).

50. Torrecillas (1997).

51. As a consequence of the catenary shape of the vault.

52. Translated as *Free-standing vaults of catenary directrix without tympanum*, Dieste (1994a).

53. The theory of elasticity assumes that a material when deformed under load returns to its original state when the load is removed.

54. In fact, the bus station in Montevideo is organised like an airport with check-in desks and departure gates, etc.

55. There are three conditions that must occur simultaneously for water to penetrate the building envelope. There must be water on the surface of the wall, there must be holes in the walls and there must be an adequate force to push the water through. Eliminate any one and there will be no water penetration.

56. From the Oxford English Dictionary.

57. Antonio Dieste is a partner in the firm Castro and Dieste.

58. This point is further discussed in chapter five.

59. All reinforced concrete is likely to crack under the full design load, concrete between the cracks makes very little contribution to the strength of the structure. Cracks may also have a detrimental effect on durability. There has been much research into crack width in reinforced concrete beams in order to minimise the size of cracks.

60. See chapter one.

61. Arup (1970), pp 265-6.

Chapter Three

62. Torrecillas (1997).

63. The Gaussian vault is named after the mathematician Karl F. Gauss (1777-1855) who described the geometry of curves. See Salvadori and Norton (1990).

64. Cosmic economy is a term described by Dieste and explained in chapter one.

65. For a given span and rise of an arch the self-weight and the compressive stress are proportional to the cross-sectional thickness of the arch. As the self-weight increases the cross-sectional area also increases and the compressive stresses remain the same.

66. Orton A. (1988).

67. The span to depth (or thickness) ratio is a typical rule of thumb used by engineers for the initial sizing of structural members. For further information see Macdonald (2001) or Orton (1988).

68. Euler buckling theory on which most design methods are based, assumes that buckling occurs due to secondary moments generated by the compressive force in the structural element. See, for example, Gere and Timoshenko (1997).

69. Ribs were actually used to support the side roof of the church of San Pedro, Durazno. However the roof was flat, predetermined by the existing plan and wall. The ribs were placed externally to maintain an uninterrupted internal soffit.

70. For a given span a separate and unique catenary curve is generated independent of the rise of the vault at its mid-point.

71. The force in a catenary is directly proportional to its rise. The greater the rise the lower the internal forces. Therefore, as the catenary shape changes along the axis of the building the internal force and,

72. hence, the deformation of each catenary segment will vary and shearing forces will develop between segments.

73. Frei Otto used soap film bubbles and fabric models to determine and develop the geometry of tensile structures.

74. Eladio Dieste in an essay *Architecture and Construction*, in Torrecillas (1997).

75. The book actually provides a review of elastic stability in general as well as its application in double-curvature vaults. An English version of this book (Dieste, 1997) is available, *Eladio Dieste 1943—1996, Calculation Methods*.

76. The finite element method allows complicated structural forms to be modelled numerically and is used extensively in the analysis of shells and other structures.

77. This would be a very expensive construction and is mentioned only to aid discussion.

78. District of Canelones.

79. Often called north light roofs in respect of their orientation to avoid direct sunlight. In the southern hemisphere these would be called south light roofs.

80. 'The tie rod is always ugly', Torroja (1967).

81. Deeper than would be necessary to deal with buckling.

82. The natural slope of a granular material is the angle to which it will naturally slump if it is not retained. Also known as the angle of internal friction.

83. The heaviest rolled sections commonly available are 200-250 mm channels.

84. Trans. Light as a brick. Piaggio (1997).

85. Wilkinson (1991).

86. Architect, Sir Norman Foster.

87. Architect, Lord Rogers.

88. Davies (1988), p 9.

89. The building cable rod generates three different load paths each dealing with a different type of load. See Orton (1988) for a fuller explanation.

90. See, for example, Brookes A. J. (1986). *Renault Building, concepts in cladding*.

91. Eladio Dieste in interview, Bayón and Gasparini (1979).

Chapter Four

92. Attributed to George Pace in Harwood E. (1998).

93. Torrecillas (1997).

94. In other texts the church is generally referred to as the Church of Atlantida.

95. 'For this building I was contracted around 1952 to design a vault. After an incredible and amusing process, which would be interesting to recount if I could do it without hurting anyone my contract was transformed into a contract for a church.' Torrecillas (1997).

96. Piaggio (1996).

97. Eduardo Dieste, in conversation with the author, Montevideo Sep. 1998.

98. Grampone J. (1996). *Eladio Dieste, Engineering Master.* Unpublished thesis, Montevideo.

99. Functional in a minimalist manner.

100. Galpón: translation, a shed, Larousse Spanish Dictionary.

101. Laity are baptised people who have not joined the religious profession.

102. Lang J. P. (1989). *Dictionary of the Liturgy.* Catholic Book Publishing, New York.

103. The mass had always been performed in Latin.

104. Harwood E. (1998).

105. Eladio Dieste at the axis of history, Marina Waismann.

106. Particularly Dieste's ideas of cosmic economy.

107. In the later reconstruction of San Pedro in Durazno, Dieste redesigned the church to merge the side aisles with the central nave.

108. In fact the net horizontal plan section is constant at all levels from floor to eaves.

109. The sacristy is a room adjacent to the altar where the priest prepares for the mass.

110. Torrecillas (1997).

111. There are many definitions of the word 'church'. In this context the church is defined as the Church of God, the community of believers.

112. Cantwell W. (1965). Design of churches and altars. *Church Building*, Oct., 16, pp 5-11.

113. Curtin et al. (1984).

Endnotes

113. See also chapter five on Durazno.
114. E. Dieste in Bayón and Gasparini (1979), p 209.
115. In 'tilt up' construction the walls are cast horizontally, usually against a flat floor slab formwork; placing of reinforcement and compaction are greatly simplified. Once the concrete has cured the wall is tilted up into a vertical position and attached in place.
116. See chapter three.
117. Dieste has used this type of detail in Gaussian vaults, see chapter three.
118. Lang J. P. (1989). *Dictionary of the Liturgy*. Catholic Book Publishing, New York, p 209.
119. Dieste went on to make extensive use of these ceramic tubes in the Church of San Juan de Avila in Spain, see chapter eight.
120. See chapter eight.
121. Piaggio (1996).
122. The Chapel at Ronchamp, Le Corbusier, Architectural press.
123. 90% of the workforce were unskilled workers from the parish supervised by a few skilled craftsmen from Dieste y Montañez.
124. Torrecillas (1997), p 163.

Chapter Five

125. The Church of Our Lady of Lourdes in Montevideo was actually started before but never completed.
126. See chapter four.
127. The apartment block was constructed after Dieste's reconstruction.
128. From Dieste's correspondence.
129. Torrecillas (1997).
130. The ascension followed the resurrection of Christ and symbolises his final return to heaven.
131. The inclination of the nave wall provides a heightened transition between the light at the top and the darker areas at the bottom.
132. Ductility is the ability to deform to a permanently altered state without cracking.
133. Consider the walls of Montevideo Shopping, see chapter seven.
134. Eladio Dieste, translated from La conciencia de la forma (The awareness of form) in Torrecillas (1997).
135. In conversation with Vittorio Vergalito, Sep. 1998.
136. From the middle of the 1960s into the 1990s there was a large community of researchers throughout the world studying many aspects of the structural performance of masonry. Codes of Practice were being developed for the design of reinforced and prestressed masonry.
137. The various methods of prestressing are described in greater detail in chapter two.
138. In other more pragmatic buildings such as warehouses and gymnasia Dieste does make use of horizontal ties.
139. Sutherland (1981), pp 31-63.
140. A view also taken later by C. T. Grimm, see chapter one.
141. Dieste (1997). *Ténica Y Subdesarollo* (Technology and underdevelopment) in Torrecillas (1997).
142. Curtin (1980), pp 41-8.
143. Diaphragm walls are wide cavity walls, in which the skins are held up to 700 mm apart by cross ribs at regular intervals, creating a large box shaped cross-section. They gained popularity in tall single storey buildings, typically sports halls and assembly buildings. Often these buildings were constructed using a conventional portal frame, clad in non load-bearing masonry. It could by shown that the increased structural section of the diaphragm wall could carry the wind loading and therefore eliminate the portal frame and, hence, generate considerable cost savings.
144. Curtin et al (1981), pp 411-17.
145. See chapter seven.
146. *Building Design*, 23 May, 1986.
147. It should be noted that Bradshaw used post-tensioned masonry in retaining walls.
148. From 1980 through to 1986 an extensive programme of research at the University of Edinburgh confirmed the ability for brickwork to be prestressed to high levels of stress and that the behaviour was comparable to equivalent prestressed concrete structures. See Walker (1987).
149. The problem of the lack of an historical perspective in the education of engineers is discussed in *The art of structural engineering*, Addis (1990).
150. In the catholic mass transubstantiation is

150. the point at which the bread of the Eucharist becomes the body and blood of Christ.

151. The apparent colour and brightness of brickwork is greatly influenced by the colour and thickness of the pointing of the bed joints.

152. The design for San Pedro was later used for a new church in Spain, see chapter eight.

153. Torrecillas (1997).

Chapter Six

154. For example, in a stool a minimum of three legs or reaction points are needed to maintain stability.

155. For example, the festival of Britain or the Torre de MontjuÔc commissioned to celebrate the 1992 Olympic Games.

156. In a Newtonian view of stability it is clearly stable otherwise it would have fallen down. This is not to say that it will always be this way if its inclination continues to increase. Great efforts are being made to preserve it in a 'leaning' state.

157. Heinle and Leonhardt (1989).

158. Torrecillas (1997).

159. Heinle and Leonhardt (1989).

160. Heinle and Leonhardt (1989).

161. In concrete construction the finished quality of the concrete surface cannot be checked until the concrete has hardened and the moulds have been stripped.

162. Wind speed and, hence, pressure fluctuate continuously. In order to provide suitable values for the design wind speed the recorded wind speed is averaged over a particular duration, in the UK this is taken as between 3 and 15 seconds. Higher average wind pressures are obtained when averaged over shortest gust duration.

163. The natural period of vibration of typical towers is considerably greater than 3 seconds and, therefore, the assumption of static loading is appropriate.

164. A more detailed explanation of the structural calculations is provided in Dieste (1997).

165. Most communication towers have to accommodate two types of transmitters, for broadcasting (the antennae) and directed signals (parabolic dishes). See Cerver (ed.) (1992).

166. This comparison is rather crude and factors such as overall height, local topography and wind conditions would have to be taken into account for a more detailed analysis.

167. Cerver (ed.) (1992).

Chapter Seven

168. In the USA there are over 4 billion ft2 of shopping mall and 10% of the population now work in malls. Beddington N. (1991). *Shopping centres*. Butterworth Architecture.

169. Speed of construction is a vital element of contemporary commercial capital projects. Faster construction times lead to reduced finance cost and sooner revenue returns. In the UK in the 1980s the term 'fast track' was used to define projects that used construction and management techniques geared towards early completion.

170. A later extension added a further 2600 m^2.

171. The span of the central vault is half that of the two side vaults and the height is much greater than is needed for purely structural reasons.

172. See chapter three.

173. The effect is weakened however by the incorporation of large air conditioning ducts either side of the mall.

174. In larger vaults with less curvature the adjustment of the bricks to follow the surface is more gradual and can be accommodated by variations in the joints between bricks.

175. like the surface formed by the strings of a violin passing over the bridge.

176. The concrete frame of the window is quite deep and will not allow a great passage of light, relative to the area behind and it is too high for views either out or in.

177. The stressing end is the position of the point of application of the prestress. The anchorage at the opposite end is often referred to as the dead end.

178. It is more common for the buttresses to be wider at the bottom and taper towards the top.

179. In order to maintain discipline in the workforce. There was a general rule that failure to arrive on time would lead to a loss of a day's pay. A rule that Pacheco normally enforced rigorously. He allowed one exception to this rule in the case of a

Endnotes

particularly wayward, free spirited worker who, although he was regularly late, nevertheless worked hard and had a greater output than most of the other workers. The unions found out of this unfairness and approached Pacheco. Fearing a potentially disastrous industrial dispute he quickly explained that as the worker was Dieste's 'son' what else could he do? (of course he wasn't) In a spirit of Latin American understanding for passion and emotion the union official sympathised with his dilemma and the dispute never progressed any further. Eduardo Dieste 1998.

180. See Alexander (1979).

181. Torrecillas (1997).

182. In the southern hemisphere from the NE direction.

183. Torrecillas (1997).

184. Patterns of events or of space.

185. It is not uncommon for designers to use their own houses as show pieces for their skills.

186. With Antonio Bonet.

187. In 1968 the children ranged from 4-17 years in age. The children would start to leave the family home in a few years.

188. Torrecillas (1997).

189. A cone is a particular form of dome the vertical section of which has straight sides.

190. Newtons' Third Law states that for every action there is an equal and opposite reaction; in other words whenever one body exerts a force on a second body the second body exerts a force on the first of the same magnitude and line of action. Conversely, if there is no force in the upper compression ring of the cone there can be no force at the base of the cone in the tension ring.

191. Trans. Our Lady of Lourdes.

Chapter Eight

192. Dieste (1997). *Ténica Y Subdesarollo* (Technology and underdevelopment) in Torrecillas (1997).

193. Dieste quoted in an article, El Corredor de Henares, by Carlos Clemente and Juan de Dios de la Hoz. Web Architecture Magazine (www.web.arch-mag.com). 7.

194. Sobejano (1992).

195. See chapter one.

196. Carlos Clemente is an award winning architect, Director of architecture for the University of Alcal· and also Director of the postgraduate course in rehabilitation and conservation.

197. In 1498 Cardinal Ximénez de Cisneros founded the University. The University had its own printing press and produced the first version of the Polyglot bible using parallel texts in Latin, Greek, Hebrew and Chaldean.

198. At least 20 other towns and cities claim to be the birthplace of Cervantes (1547-1616), however his baptism is recorded in Alcal·.

199. Originally El Corredor de Henares was part of the Diocese of Madrid.

200. Clemente C. and de Dios de la Hoz J. (1996). Tres Nuevos Templos en la Diocesis de Alcal·. Ars Sacra, May, pp 17-31.

201. Community buildings and meeting rooms, etc.

202. Carlos Clemente had extensive experience of urban restoration and has won a number of international awards. He was also Director of the Spanish Institute of Architects (1992-94).

203. System building using prefabricated standardised components for example.

204. In the 1960s, system building was proposed for churches but the idea did not take off. Catt J. (1965) Design entry for a system built church. Church Builder. 16.

205. The balustrades at Atlantida were constructed as vertical uprights of single bricks laid on their ends and have broken in places.

206. The term 'pinned' refers to a condition where the connections between two elements allows relative rotation between them. In the case of the wall at Atlantida, the wall does not undulate at the connection with the foundation, it is simply the thickness of the wall. In comparison with its height the connection between the wall and the foundation is relatively flexible and will rotate slightly. The bending moment at a pinned base is zero.

207. A fixed base implies a considerable degree of rigidity between the wall and the foundation. In San Juan the undulations of the base of the wall will impart much more rigidity to the structure than the straight connection between wall and foundation at Atlantida. The greater rigidity resists

208. In reality the deformations are small, impossible to perceive with the human eye. In the diagram, they are sketched to an exaggerated scale.

209. The position in a structure where the bending moment changes sense is known as the point of contra-flexure, At the point of contra-flexure there is no bending moment. A simple model for a point of contra-flexure can be obtained using a flexible plastic ruler. Hold the ruler firmly at each end, rotate both ends of the ruler in the same direction, either clockwise or anti-clockwise, while maintaining a firm grasp. The ruler should deform to a flattened S shape showing both convex and concave curvature. The point of contra-flexure occurs at the transition between concavity and convexity, where there is no curvature.

rotation and, hence, develops bending moments at the junction between wall and foundation.

210. See *Structures*, chapter 9, by Schodek (1980) for further information.

211. See chapter five.

212. The diaphragm wall has an I-shaped cross-section.

213. *Costes + Datos de Edification*, 16, July 1998. The journal monitors the construction costs of recent buildings in Spain.

214. The exchange rate for Euros to Sterling in April 2000 was around 1·6 Euros to £1. The project cost approximately £367 000.

215. Trans. 'Our Lady of the Rosary'.

216. Author's note. When trying to find the church the author asked directions using both the address and photographs of the church and encountered a number of inhabitants who did not know of the church or its location.

217. There are 14 Stations of the Cross and they represent the stages of Christ's death and resurrection and form an important part of the Easter Celebrations of the Catholic Church.

218. Translation of *Camino De Los Estudiantes*, from Torrecillas (1997), is the *Student Lane*, which does not seem to convey the spirit of the idea behind the project.

219. In Uruguay snow fall is almost non-existent and is not normally considered a necessary factor in structural design.

220. The most appropriate definition of arbour in the OED is 'a shady retreat'.

221. Baldwin R. (1998). Challenging brickwork. The Brick Bulletin, Autumn, The Brick Development Association.

222. Carlos Clemente, in correspondence with the author, Feb. 2000.

223. Clemente and de Dios de la Hoz were responsible for a further four churches in the diocese, built using more conventional methods.

224. Trans. Institute for Structural Design and Research.

225. See chapter four.

Conclusion

226. In the early seventies Dieste was involved in politics for a short period.

227. He was selected as representing Uruguyan Architecture in a review of Latin American Architecture sponsored by Unesco.

228. Dieste has described the design of a simple prestressing jack where he took care over shaping the form over and above that which was needed for purely functional needs.

229. Fernández Ordóñez and Navarro Vera (1999).

230. Holgate (1986), p 264.

231. Bayón and Gasparini (1979), p 212.

232. Huber and Steinegger (eds) (1971).

Selected List of Works

1947
Berlinghieri House, Punta Ballena, 6 m span vaults, Uruguay
Architect: Antonio Bonet.

1955
Warehouse for ANCAP, 8 m span vaults, Capurro, Montevideo, Uruguay,

1955
Furgoni Warehouse, 22 m span vaults, 3 500 m², Montevideo, Uruguay

1956
Warehouse for newspaper El Pais, 22 m span vaults, 2 200 m², Montevideo, Uruguay

1960
The Church of Jesus Christ the Worker, curved walls, Gaussian vaults, 500 m², Atlantida, Uruguay

1962
TEM Warehouse, 43 m span Gaussian Vaults, 8 200 m², Montevideo
Architects: Clerck and Guerra.

Unfinished
Nuestra Señora de Lourdes, parish church, 1 000 m², Malvin, Montevideo

1968
Casa Dieste, private house, 350 m², Punta Gorda, Montevideo

1970
The Church of San Pedro, 32 m span pre-stressed walls, 600 m², Durazno, Uruguay
Architect: A. Castro and engineer, R. Romero.

1971
Fruit Growers Pavilion, 35 m span Gaussian vaults, 35 000 m², Porto Alegre, Brazil
Architects: CM Fayet and C Araujo.

1972
Municipal Bus Station, 15 m cantilever vaults, 900 m², Turlit, Salto, Uruguay
Architect: N. Minutti.

1973
Market, 7 m span vaults, 150 000 m², Rio de Janero, Brasil

1974
Municipal Bus Station, 13.5 m double cantilever vaults, 1 300 m², Salto, Uruguay
Architect: N. Minutti

1976
Petrol Station for Barbieri and Leggire, 9 m span double cantilever canopy, 1 500 m²,Salto, Architect: N. Minutti

1976
Parador Ayui, restaurant, 520 m², Salto
Architect: N. Minutti.

1977
Warehouse for Agronindustria Domingo Massaro, 16.5m cantilever vaults, 10 000 m², S.A. Canelones, Uruguay

1979
Warehouse for Julio Herrera and Obes, 50 m span Gaussian Vaults, 4 000 m² Montevideo, Uruguay

1979
Water Towers for Refrescos del Norte, Salto, Uruguay, 24 m high
Engineer: R. Romero.

1980
Rowing Club of Salto, gymnasium, 500 m², Salto, Uruguay
Architect: Ambrosoni.

1990
Fagar Cola Bottling Plant, 13m cantilever vaults, 3 900 m², Colonia, Uruguay

1994
ADF Wool Depot, 40 m span Gaussian Vaults, 10 000 m², Canelones, Uruguay

1997
Storage Silo, 45 m span, 6 800 m² , Nueva Palmira, Colonia, Uruguay

1985
Montevideo Shopping Center, Gaussian and Barrel vaults, undulating walls, 16 m span, 10 000 m² , Montevideo
Architects: Studio Gómez Platero-López Rey

1986
Television Communications Tower, 66 m high, Maldonado, Uruguay.

1996
San Juan de Avila, parish church, 500 m² , Alcalá de Henares, Spain
Architects: Carlos Clemente and Juan de Dios de la Hoz.

1997
Madre del Rosario, parish church, 540 m², Mejorado del Campo, Spain
Architects: Carlos Clemente and Juan de Dios de la Hoz and José Luis de la Quintana.

1997
La Sagrada Familia, parish church, 500 m², Torrjon de Ardoz, Spain
Architects: Carlos Clemente, Juan de Dios de la Hoz and Ana Marin.

Under construction
Student promenade, University of Alcalá, 52 barrel vaults, each 30 m long and 3 large brick cones, Alcalá, Spain
Architects: Carlos Clemente, and Ana Marin.

153

Bibliography

Addis W. (1990). *The art of structural engineering: the nature of theory and design.* Ellis Harwood.

Alexander C. (1979). *The timeless way of building.* Oxford University Press.

Arup O. (1970). The potential of prestressed concrete. Concrete, June.

Baldwin R. (1998). Challenging brickwork. *The Brick Bulletin,* Autumn, The Brick Development Association.

Bayón D. and Gasparini P. (1979). *The changing shape of Latin American architecture.* John Wiley and Sons.

Brookes A. J. (1986). *Renault Building, concepts in cladding.* Construction Press.

Bullrich F. (1969). *New direction in Latin American architecture.* Studio Vista Limited, London

Castedo L. (1969). *A history of Latin American art and architecture.* The Pall Mall Press, London.

Cerver F. A. (ed.) (1992). *Communication towers.* Edicones Atrium, Barcelona.

Curtin W. G. (1980). Brick diaphragm walls - Development, application design and future development. *The Structural Engineer,* 58A, 2, Feb.

Curtin W. G., Shaw G., Beck J. K. and Bray W. A. (1981). Diaphragm walls. *Architects Journal,* 26 Aug.

Curtin W. G., Shaw G., Beck J. K. and Bray W. A. (1984). *Structural Masonry Detailers' Manual.* Granada Publishing, London.

Davies C. (1988). *High Tech architecture.* Thames and Hudson, London.

de Dios de la Hoz Martines J. and San Román C. C. (1998). La Construcción con Cerámica Armada. Iglesia de San Juan de Avila, en Alcalá de Henares/España (Building with reinforced ceramic, San Juan de Avila Church, in Alcalá de Henares/Spain). *Informes de la Construcció,* 49, 453, Feb., 41-53.

Diehl K. L. (1990). Eladio Dieste, Revolution im Ziegelbau (A revolution in brick architecture). *Ziegelindustrie International,* 6/90.

Diehl K. L. (1991). Bauen mit bewehrten Zieglschalen (Reinforced brick structures). *Ziegelindustrie International,* 10/91.

Dieste E. (1992). Some reflections on architecture and construction. *Perspecta,* 27.

Dieste E. (1994a). *Cascaras autoportantes de directriz catenaria sin timpanos* (Free-standing vaults of catenary directrix without tympanums). Ediciones De La Banda Oriental, Montevideo.

Dieste E. (1994b). *Pandeo de Laminas de Doble Curvature* (Deflection in double-curvature vaults). Ediciones De La Banda Oriental, Montevideo.

Dieste E. (1997). *Eladio Dieste, 1943-1996, Métodos de cálculo.* Junta de Andalucia. English version *Eladio Dieste 1943-1996, Calculation methods.*

Dieste E. (1998). Un Continente sumergido (A submerged continent). *AV Monografías,* 69-70 April, 164-167.

Fernández Ordóñez J. and Navarro Vera J. (1999). *Eduardo Torroja - Engineer.* Ediciones Pronaos, Madrid.

Gere J. M. and Timoshenko S. P. (1997). *Mechanics of materials.* PWS Publishing Company.

Grimm C. T. (1997). Research and Innovation. *Masonry International.* 2, 2, 36-37.

Heinle E. and Leonhardt F. (1989). *Towers: a historical survey.* Butterworth Architecture.

Harwood E. (1998). Liturgy and Architecture: The development of the centralised Eucharistic space. *Twentieth Century Architecture.* 3, 49-74.

Holgate A. (1986). *The art of structural design.* Clarendon Press, Oxford.

Huber B. and Steinegger J.C. (eds) (1971). *Jean Prouvé - Prefabrication: structures and elements.* Pall Mall Press Ltd. London.

Macdonald A. J. (2000). *The engineer's contribution to contemporary architecture - Anthony Hunt.* Thomas Telford, London.

Mainstone R. (1975). *Developments in structural form.* Allen Lane, London.

Mark R. (ed.) (1993). *Architectural technology.* The MIT Press, Massachusetts.

Ochshorn J. (1999). *Brick, encyclopedia of twentieth century architecture.* Fitzroy Dearborn.

Orton A. (1988). *The way we build now.* E &FN Spon.

Paduart A. (1966). *Shell roof analysis.* GR Books, London.

Piaggio J. M. (1996). Eladio Dieste, L'ingegno e l'architettura (Eladio Dieste, the engineer and the architect). *Costruire in Laterizio,* 52-53/96, 156 - 179.

Piaggio J. M. (1997). *Leggero Come un Mattone* (Light as a brick). Industria laterizi Giavarini.

Salvadori M. and Norton W. W. (1990). *Why buildings stand up*. USA, ISBN 0-393-30676-3.

Schodek D. L. (1980). *Structures*. Prentice-Hall, New Jersey.

Sinha B. P. and Pedreschi R. F. (1992). *Reinforced and prestressed brickwork*. The Institution of Engineers, India.

Sobejano E. (1992). Wall and skin: Contemporary architecture in Spain. *Daidalos*, 43.

Sutherland R. J. M. (1981). Brick and block masonry in engineering. *Proc. Instn Civ. Engrs*, Part 1, 70, Feb.

Strike J. (1991). *Construction into design*. Butterworth Architecture, Oxford.

Torrecillas A. J. (1997). *Eladio Dieste, 1943-1996*. Junta de Andalucia.

Torroja E. (1967). *Philosophy of structures*. University of California Press, Berkley.

Walker P. (1987) *A study of the behaviour of partially prestressed brick work beams*. PHD thesis. University of Edinburgh.

Wilkinson C. (1991). *Supershed*. Butterworth Architecture.

Index

Addis, W.	22
ADF Wool Warehouse	53, 54, 119
Agroindustry Massaro	36–38, 40
Alcalá	122–124
Alcalá, University	28, 122, 135
Alexander, C.	112, 113
ANCAP	28
Arch	30, 31
Arup Associates	30
Arup, O.	43, 142
Atlantida, Church of	66–79, 83, 86, 98, 106, 109, 124, 127, 142
Baker, L.	21
Baptism	71
Barrel Vaults	31, 43, 116
Bayón, M.	122
Bell tower	96, 97
Berlinghieri House	14, 16, 17, 28, 112, 114
Bonet, A.	16, 28, 112
Bradshaw, Buckton and Tonge	90
Brick Institute of America	25
Brunel, I. K.	24
Buckling in vaults	50
Bus Terminus, Salto	38, 39
Camino de los Estudiantes	123, 135–139
Candela, F.	22, 23, 30, 46, 73, 143
Canelones	53
Casa Dieste	112–117
Castle of Molina de Aragón	122
Castro, A.	83
Catenary arch	30
Catholic church	68, 71, 76
Cervantes, M.	123
City of God	123
Clemente, C.	123, 124, 138
Club Remeros	119
CN tower	102
Collserola tower	102
Concrete shell	16
Corredor de Henares	123
Cosmic economy	24, 46, 144
Cristiani and Neilson	16
Curtin, W.	73, 90, 128
Davies, C.	63
de Dios la Hoz, J.	123, 124, 139
Diaphragm Wall	90, 131
Diehl, K.	99
Dieste, Antonio	42
Dieste, Eduardo,	21
Dieste, Eladio senior	14
Dieste, Rafael	14
Dieste y Montañez	17, 20, 28, 50, 53, 106, 137
Dome	28
Don Quixote	123
Durazno	82, 106
El Pais	28
Emley Moor Tower	102
Fagar Cola	39–41, 101
Fathy, H.	21
Fayet and Araújo	53
Fifth Congress of the City of Knowledge	122
Fleetguard Factory	63
Foster, N.	102
Freyssinet, E.	30, 42, 43, 62, 143
Friedham, E.	16, 139
Galeria des Machines	30
Garcia de Zunga, E.	14
Gaussian vaults	46–63, 68, 106
Generation of '45	14
Gomez, Platero-Lopéz, Rey	107
Gothic	62
Grimm, C.T.	25
Guggenheim Museum	22
Guidice, A.	67, 68
Hannover University	139
Harris and Sutherland	90
Heinle, E.	96, 98
Henares river	123
Hernandez, A	19
High Tech	30, 63
Hochschule, Berlin	139
Imperial War Museum	30
Insitut fur Tragwerksenturf und Bausweisenforchung	139
Institution of Civil Engineers	90
Isler, H.	16
Jesus Christ the worker (church)	66
Kahn, L.	21
Landmark Entertainment Centre	138
La Gente Sencilla	20, 143
Larrembebere, G.	21
Last Supper	71
Leaning Tower of Pisa	96
Le Corbusier	16, 78, 79
Leonhardt, F.	96, 98
Liturgical Movement	68, 85
Load test	52
Madrid	123
Maillart, R.	30
Mainstone, R.	30
Mejorado del Campo	131–134
Millennium Dome	22
Moneo, R.	122, 123
Montañez, E.	17, 18, 20
Montevideo Shopping Centre	106–112
Montevideo Warehouse	54–57, 75, 108
Munoz de Pablos, C.	125
Nervi, P.L.	22, 143
Newton's Third Law	118
Niemeyer, O.	73

Nuestra Madre del Rosario (church)	131–134
Nuestra Senora de Belen (church)	139
Nuestra Senora de Lourdes (church)	118–119
Nueva Palmira (silo)	19, 60–62
Ochshorn, J.	21, 123
Orton, A.	48
Otto, F.	16, 49
Our Lady of Lourdes (church)	90, 124
Ove Arup and Partners	102
Pacheco, V.	19, 112
Pantheon	42
Parador Ayui	117–118
Pedreschi, R.	25
Piaggio, J.M.	62, 67, 79
Ponds Forge Sports Centre	30
Pope John XXIII	68
Pope Pius XII	68
Porto Alegre Market	19, 53, 54
Post-tensioning	90
Prestressing	34, 37, 42, 87–91
Prouvé, J.	143, 144
Pseudo-rationalism	66
Renault Distribution Centre	63
Rice, P.	63
Rio de Janerio Market	19
Romero, R.	83
Ronalds, T.	138
Ronchamp (church)	79
Royal Glass Workshop of La Granja	125
Sagrada Familia (church)	135
Salvadori, M.	32
San Juan De Avila (church)	124–131
San Pedro (church)	76–93, 124, 131, 132, 134
Santa Cruz (church)	139
Scientific Revolution	30
Seagull	40–42
Second Vatican Council	68
Sinha, B.P.	25
Sobejano, E.	122
Solsiro silo	59
Spanish Civil War	14
St Hedwig's Church	25
St Pancras Station	30
Striew, H.	20
Sutherland, J.	90
Sydney Opera House	30
Taller Francisco Suttener	20
Taller Garcia	14
Television Tower Maldonado	102, 103
TEM Warehouse	49
Torrejon de Ardoz	135
Torres-Garcia	14
Torroja, E.	22, 23, 30, 46, 142, 143
Tympanum	28, 32, 49
UNESCO	123
Universal Constructivism	14
Utzorn, J.	30
Vergalito, V.	19, 87, 132
Viermond SA	16, 68, 111
Waismann, M.	68, 78
Waterloo Rail Terminal	30
Water Towers	97, 98
Wilkinson, C.	62, 63